Co-operative Approaches
to Sustainable Agriculture

ORGANISATION FOR ECONOMIC CO-OPERATION AND DEVELOPMENT

ORGANISATION FOR ECONOMIC CO-OPERATION AND DEVELOPMENT

Pursuant to Article 1 of the Convention signed in Paris on 14th December 1960, and which came into force on 30th September 1961, the Organisation for Economic Co-operation and Development (OECD) shall promote policies designed:

- to achieve the highest sustainable economic growth and employment and a rising standard of living in Member countries, while maintaining financial stability, and thus to contribute to the development of the world economy;
- to contribute to sound economic expansion in Member as well as non-member countries in the process of economic development; and
- to contribute to the expansion of world trade on a multilateral, non-discriminatory basis in accordance with international obligations.

The original Member countries of the OECD are Austria, Belgium, Canada, Denmark, France, Germany, Greece, Iceland, Ireland, Italy, Luxembourg, the Netherlands, Norway, Portugal, Spain, Sweden, Switzerland, Turkey, the United Kingdom and the United States. The following countries became Members subsequently through accession at the dates indicated hereafter: Japan (28th April 1964), Finland (28th January 1969), Australia (7th June 1971), New Zealand (29th May 1973), Mexico (18th May 1994), the Czech Republic (21st December 1995), Hungary (7th May 1996), Poland (22nd November 1996) and Korea (12th December 1996). The Commission of the European Communities takes part in the work of the OECD (Article 13 of the OECD Convention).

Publié en français sous le titre :
ACTIONS CONCERTÉES EN FAVEUR DE L'AGRICULTURE DURABLE

FOREWORD

Across the OECD, farmers are voluntarily forming community-based associations to work for a better environment.

How can one account for this phenomenon and how can it be encouraged? To what extent can such group-based voluntary action substitute for, or at least complement, existing agri-environmental policy measures? In attempting to answer these questions, this study looks at the experiences of four OECD Member countries: Australia, Canada, The Netherlands and New Zealand. It also provides a framework for understanding the role that such groups may play in promoting a more sustainable agriculture. By examining the relationships between these groups and different levels of government, particularly as regards measures taken to encourage the formation and continual survival of these groups, the study seeks to identify the conditions under which such groups seem to work most effectively, and the types of issues to which they may be best suited. In so doing it provides a new perspective on the role of voluntary, collective action in finding local solutions to local environmental issues in agriculture.

Governments are increasingly finding that for many categories of environmental problems, implementing policies through responsible groups can be more effective or less costly than those aimed at individuals. Such approaches can be more flexible than alternatives, allowing greater scope for experimentation, and can minimise budget costs and the need for direct intervention by central government. Moreover, they may be the most appropriate for dealing with the many environmental issues that are predominantly local in nature. In agriculture, these environmental issues include the effects on local communities of wind-borne soil erosion, exotic pests and pesticides, and the provision of landscape amenities.

The motivation for farmers to form or join landcare groups appears to relate to protecting the value of their farm assets and avoiding regulations that they would regard as burdensome. The farmers believe that, by taking their own initiative, they would be more likely than some outside authority to achieve satisfactory, locally acceptable solutions to externality problems.

The study also shows that there is great diversity in the specific initiatives being undertaken by landcare groups. Some have formed mainly to provide a forum to exchange information; others have taken more active initiatives, financed at least in part by their members. The most common activity has been to work together in preparing farm plans. These farm plans usually take a "whole farm approach", encouraging farmers to consider all the environmental, economic and sociological factors that bear on the sustainability of their enterprises. Typically, outside advisors are enlisted to help design and use the plans. While landcare groups tend to focus on local issues, they usually broaden their perspectives over time, and begin to look at sustainable land management from an "integrated systems" perspective.

How effective farmer-led initiatives have been in tackling environmental issues is difficult to answer at this stage. However, such farms tend to have farm management plans and adopt soil conservation practices sooner than non-members. As more farmers maintain farm plans, and begin to monitor their performance against indicators of economic, physical and biological conditions, more thorough evaluations will be possible.

The study was written by Ronald Steenblik, based in part on material provided by the four Member countries studied. Useful guidance was provided by Wilfrid Legg. It is the result of a study carried out by the Joint Working Party of the Committee for Agriculture and the Environment Policy Committee, and forms part of a larger effort to explore the linkages between agricultural policies and the environment, and to identify ways in which governments can address environmental objectives in agriculture with minimal resource cost to the economy and the fewest trade distortions. The two parent committees approved the report in 1997 and agreed that it be published under the responsibility of the Secretary-General of the OECD.

TABLE OF CONTENTS

TABLES

FIGURES

BOXES

INTRODUCTION

"It is a mistake to limit collective action to State action; ..."
Kenneth J. Arrow (1977)

In several Member countries of the OECD an interesting phenomenon is taking place. Farmers, together with other stakeholders, are forming voluntary associations to improve their local environments. These spontaneously formed, community-based associations are known by different names: landcare groups, conservation clubs, environmental co-operatives and many others. Despite differences in their origins, governance, methods and types of activities in which they become involved, they appear to share one very important characteristic in common: a desire to achieve the basic goals of sustainable agriculture through a participative process that reflects the values of their own communities.

The emergence of voluntary, farmer-led groups has not gone unnoticed by central governments. Nevertheless, the response to them has varied considerably. In several countries where such groups are active in agriculture, government policy has been encouraging but cautious – some are taking a "wait and see" attitude, others have decided to experiment with these groups to determine whether they can live up to their promise. By contrast, in Australia, where voluntary community-based ("landcare") groups have been active for more than a decade, the federal and State governments have developed major programmes to complement them.

Such experimentation with and promotion of community-based approaches to sustainable development is by no means confined to the agricultural sector, nor is it particularly new. The idea of community control over community problems has been a recurrent ideal of political philosophers since the time of Aristotle. Nowadays there are also some very pragmatic reasons why many policy-makers are attracted to the idea of delegating more responsibility for implementing and monitoring environmental policy to private groups: such approaches can be flexible, allow for experimentation, and minimise budget costs and the need for direct intervention by central government. But perhaps more importantly they may prove the most appropriate scale for dealing with the many agri-environmental issues that are predominantly local in nature.

The purpose of this document is threefold. First, it seeks to better understand the current nature, scope and activities of the numerous voluntary, farmer-led groups that have formed in several OECD countries to promote sustainable agriculture and to take on greater responsibility for their local environment. Second, by examining the relationships between these groups and different levels of government, particularly as regards measures taken to encourage the formation and continual survival of these groups (including the general agri-environmental policy context), the document seeks to identify the conditions under which such groups seem to work most effectively. Third, the document considers the potential roles that these groups might play in implementing and perhaps even defining the sustainable agriculture agenda. In particular it asks: to what extent can group-based voluntary action or self-regulation in the agricultural sector substitute for, or at least complement, existing agri-environmental policy measures? This question concerns the long-term viability of the groups as well as problems of accountability and cost-effectiveness.

The organisation of the paper is as follows: Chapter 2 looks at voluntary, co-operative approaches from a public choice perspective and attempts to identify some of the situational characteristics that are likely to have a bearing on their suitability. Then follows a series of similarly structured chapters, each describing a sample of voluntary, co-operative initiatives to promote more sustainable agriculture in four Member countries: Australia, Canada, The Netherlands and New Zealand. Given its limited focus, the paper does not purport to cover the full range of voluntary, co-operative approaches being taken in the OECD or even in the particular countries examined, but to report on a sample of initiatives that represent the range of programmes under way. (In the United States, for example, community-led groups are being formed in many areas to address water-quality problems at the catchment or watershed level (see, *e.g.*, GAO, 1995); Annex describes one recent initiative in New York State). Because of the comparative wealth of material on Australia's community landcare groups, which have been operating for more years than similar groups in the other countries, the chapter on that country is considerably longer than those for the others. Finally, in Chapter 7, some policy conclusions are drawn.

VOLUNTARY GROUP ACTION
IN THE PUBLIC INTEREST: ISSUES

The legislator [should] follow the spirit of the nation, when doing so is not contrary to the principles of government, for we do nothing better than what we do freely and by following our natural genius."
Charles Louis Montesquieu (1748)

This study is concerned primarily with voluntary, co-operative initiatives to promote more sustainable agriculture. In particular it seeks answers to the following questions:

- To what sorts of environmental problems are voluntary group action best suited?

- How do characteristics of different environmental issues; benefits and costs of abatement; and farmer's attitudes towards the environment affect their willingness to participate in voluntary groups?

- What steps do policy makers need to take to foster the development of such groups?

- What are the conditions under which a voluntary compliance regime may be superior to a mandatory compliance regime?

- To what extent is the success of such groups affected by the agricultural and environmental situation and policy context in which they operate?

These are questions on which an empirical examination of actual experiences can shed much light. They are also questions that have been considered in a general way in the literature of environmental economics, public choice theory, and group dynamics, which is drawn upon in this chapter. As a starting point, the following section looks at the place that voluntary group initiatives hold in the wider set of institutional arrangements within which government policies are formed and implemented.[1]

VOLUNTARY, CO-OPERATIVE APPROACHES: A PUBLIC CHOICE PERSPECTIVE

Environmental problems stem from the behaviour of individuals or organisations, and in trying to deal with them governments historically have sought to influence such behaviour through measures addressed to the individual agents responsible. This they have done with policy measures involving degrees of compulsion ranging from none to strict. Programmes to encourage farmers to adopt soil-conservation practices, for example, have generally not been compulsory; taxes on fertilisers and pesticides have been. Individuals need not always be the primary agents of environmental policy, however. And governments are increasingly finding that for many categories of environmental problems, implementing policies through organised groups can be more effective or less costly than through individual agents.

Table 1 uses these two dimensions, degree of compulsion and implementing agent, to provide a framework for illustrating the range of structures within which governments can implement policies and programmes. Generally, the power exercised by *government* is greatest when it uses a top-down process (*i.e.*, at the left-hand side of the table). This is because compulsory restrictions by their very nature must be universally applied, requiring extensive monitoring and enforcement of compliance. At the other extreme (the right-hand side of the table), power is either allowed to reside with, or delegated to, the target group or individuals. Between these two extremes lies various forms of semi-compulsory or co-operative arrangements, involving consultation and co-operation – *i.e.*, power sharing. Within such arrangements, the government will often also make appeals for voluntary action,[2] in an attempt to change people's attitudes and ultimately their behaviour (or to encourage people to persuade each other to change), generally by supplying information so that they can make better informed decisions.

As with policies implemented through individuals, policies implemented through groups can range from those based on coercion to those based on encouragement and persuasion. One common form of institutional arrangement is the so-called "voluntary accord" (VA), or policy covenant, between a government and a target group, usually an industry association. Such VAs have proliferated in recent years, particularly in the chemical and energy sectors (see, *e.g.*, Potier, 1994; Masuhr, 1994; Solsbery and Wiederkehr, 1995), though they also include several in agriculture, such as the UK's VA with the plastic films industry to collect and recycle polythene waste from farms (Woodall, 1996 and 1997).[3]

Many VAs are forged through negotiation and might therefore be more accurately characterised as "semi-compulsory". Often the targeted groups will have entered into a VA in response to a threat by the government that it will take

Table 1. **Institutional arrangements for policy**

	GOVERNMENT					
Method of ensuring participation:	Coersion	Pressure	Persuasion	Encouragement	Facilitation	No involvement
Concrete action:	Informs	Consults	Co-operates	Advises	Monitors	Observes
Implementation responsibility of:	Examples:					
Each individual or firm	Government sets an environmental standard with little or no prior consultation with affected firms.	Government holds public hearings before setting a standard.	(Rare, because of high transaction costs.)	Government establishes "best practice" guidelines.	Government provides standard format for firms to report on activities.	Individuals engage in unreported voluntary activities.
			Voluntary, co-operative action by industry			
Groups of individuals or firms (collectively)	Government sets total emission quota for industry, which then allocates firm-level quotas.	Industry negotiates a formal covenant with Government obliging the former to achieve a specified level of performance.	Government and industry agree to share responsibility in implementing programme.	Government provides advisors to work with voluntary groups within industry.	Association collects information on members' activities and reports to Government.	Groups engage in unreported voluntary activities (e.g., establishing code of best practice).
	INDUSTRY					
Concrete action:	Is informed	Advises	Co-operates	Consults	Informs	Self regulates
Legal obligation:	Compulsory	Semi-compulsory		Co-operative		Non-compulsory

tougher action if informal agreement on the basic objectives – usually an undertaking on the part of the target group to reduce emissions by a specified amount or to change its behaviour in a certain way – is not reached. This threat thus provides, in effect, an *ex post* sanction (Glachant, 1994). The legal character of VAs is typically vague. Indeed, courts have sometimes found it difficult to determine whether a particular agreement should be regarded as a form of social regulation or a contract under private law (WRR, 1992). Consequently, the trend in several countries has been to provide for the possibility, or require, that new VAs be formalised in public law (BNA, 1995).

The role of truly *voluntary* group action in pursuit of goals that are consistent with public policy objectives – the main focus of this study – is potentially greater than that of VAs, but it is less well understood. The phenomenon of institutionalised co-operation is not entirely new, of course. The oldest democratic institutions in The Netherlands, the Water Boards (*Waterschaap*), for example, trace their origins to private initiatives by farmers and land-owners in the Middle Ages to build polders[4] and to collectively manage the supply and drainage of water. More recently, private groups have formed more or less spontaneously to manage common-property resources (such as fish[5] or wildlife), to patrol neighbourhoods in an effort to reduce crime, and of course to perform a wide range of charitable services.

The propensity to form small-scale institutions to address communal problems varies across cultures and over time, naturally. But this tendency – what Machiavelli called *virtu civile* ("civic virtue") – once in place, is incredibly durable (Lemann, 1996). Perhaps sensing this, governments have sometimes tried to propagate, or at least nourish, the growth of civic institutions. Any government's involvement with such institutions must necessarily differ in extent and nature from what it does normally, however, since it can take little or no part in weighing up the interests of the different parties (WRR, 1992). That task is usually taken on by private groups, or society at large. Rather, its function should be generally facilitating, and can be as minimal as providing seed money, supporting basic research, arbitrating disputes, and serving as a conduit for the collection and dissemination of information.

GENERAL STRENGTHS AND WEAKNESSES OF USING VOLUNTARY GROUPS AS AGENTS OF PUBLIC POLICY

There has been much analysis in recent years of the strengths and weakness of devolving responsibility for resource management to groups, especially in the context of VAs. Similarly, there is a vigorous debate underway, particularly in North America, over the effectiveness of voluntary programmes in reducing the pollution from and enhancing the environmental benefits of agriculture (see, *e.g.*, Anders Norton, *et al.*, 1994). The following paragraphs summarise the main points from both topics that are relevant to voluntary, co-operative approaches to sustainable land management in agriculture.

Allocative efficiency

One of the general advantages of the co-operative management of natural resources most often cited concerns allocative efficiency. Compared with policy approaches that impose uniform restrictions on individual decision makers, approaches that allow individuals or firms to expend different levels of effort are more apt to attain a given objective (such as reduction of pollution or conservation of a common-property resource) at lower cost (Baumol and Oates, 1988). This feature is shared also with pollution charges and tradable quotas, of course. For practical reasons, however, generally linked to complexity, many environmental problems are not amenable to such market-based approaches.

Glachant (1994) notes that inter-firm bargaining over the means of achieving environmental objectives can improve allocative efficiency, but these gains must be set against the costs of negotiation. From this it follows that the smaller the number of negotiating parties, the lower the negotiating costs. Moreover, since such costs are generated by the opportunistic behaviour of rival firms (or individuals), co-operative approaches can be expected to be more efficient in industries where such behaviour is limited by the pre-existing level of trust enjoyed between the firms (or individuals), especially if they are already accustomed to working together. This point also has a bearing on monitoring and enforcement costs, as noted below.

External economies

A sometimes over-looked advantage of co-operative approaches is that they can generate positive external economies for group members. These economies relate to the collection and sharing of information and expertise, the formation and reinforcement of expectations, the promotion of a climate of innovation, and the exploitation of economies of scale.

Up to a point, collecting and sharing information can be organised at lower cost by a group than if the same information were collected by each member individually. The crucial role of institutions – *i.e.*, formalised groups – in this context is to "organise, process, and store the essential information required to co-ordinate human behaviour" (Runge, 1984). In addition, the *act* of sharing information can itself contribute to group cohesion, by fostering and maintaining an atmosphere of trust and co-operation, leading to outcomes that are superior to those that would result from non-co-operative behaviour. Institutions can help limit non-co-operative behaviour, in particular by providing assurance of fair-mindedness and making individual contribution to a public good more attractive than a free ride (Runge, 1984).

Frequent contact among practitioners, taking place within an atmosphere of trust, can also act as a spur to innovation and mutual learning. Few have described this phenomenon as well as Alfred Marshall (1952 [1920], p. 225):

"Good work is rightly appreciated, inventions and improvements in machinery, in processes and the general organisation of the business have their merits promptly discussed: if one man starts a new idea, it is taken up by others and combined with suggestions of their own; and thus it becomes the source of further new ideas." The more recently published accounts of some of the voluntary farmer groups described in Chapters 3 through 6 of this paper often echo this observation.

Finally, co-operative groups can benefit from economies of scale in the purchase of inputs or even resource-conserving equipment and infrastructure. Loehman and Dinar (1994), for example, simulated an irrigation externality problem in the Central Valley of California to demonstrate that significant gains could be reaped through a co-ordinated strategy that took advantage of economies of scale in drainage abatement and treatment. Other examples may be found in Chapters 3 and 5 of this paper.

Costs of administration, monitoring and enforcement

The assertion is often made that voluntary regimes lead to lower monitoring and enforcement costs borne by the government than do mandatory ones.[6] (The issue of whether for a given level of performance *society* incurs higher or lower costs of enforcement is rarely examined.) This claim is often supported simply by tautological reasoning: enforcing a voluntary action invalidates it, by definition. Hanna (1995) offers a more persuasive argument, based on what she calls "the transaction costs" of government policy.

As asserted by Hanna (1995), the transaction costs of policy will vary for each task that can be shared between government and resource users, and these transaction costs will in turn depend over the long run on the extent of user participation. During the stages of problem identification and policy design, transaction costs are minimised by a top-down approach – *i.e.*, one that avoids spending time and resources in co-ordination, information dissemination and conflict resolution. However, such an approach creates uncertainty in the minds of the users as to the goals of the process, encouraging short-term actions at the expense of long-term sustainability. By contrast, a bottom-up approach, involving extensive participation by users, gives them a stake in the outcome and reduces uncertainty about process goals. Users are more likely to comply with regulations, and to adopt a stewardship ethic, when they understand and endorse the policy goals, and have some assurance of control over outcomes.

Working through co-operative groups, as opposed to individuals (or non-co-operative groups), provides added scope for control and enforcement. Group leaders can appeal to members' loyalties and apply peer pressure when that is not enough (OECD, 1997c). When the number of agents is large and diffuse this can be an advantage to governments, particularly if the leaders of the target group or groups have made a commitment to some agreed goal or mode of behaviour. In most societies, cheating one's fellow group members is regarded with greater opprobrium than cheating the government.

Governments can rarely avoid incurring at least some costs of monitoring and enforcement, however. In the case of VAs, for example, while initial responsibility for monitoring and enforcement may fall upon a private group rather than on the government, the government normally remains ultimately accountable for the group's performance. It may still need to incur expenses in monitoring the group's performance and, depending on the nature of the agreement, prosecute rene-gade violators.

Similar concerns apply to non-compulsory measures. Public appeals to people's sense of altruism, civic duty or simple self-interest that are intended to elicit voluntary actions or changes in behaviour are usually seen as being more effective in the long run than in the short run. Implicit is the expectation or hope that particular modes of behaviour will become the norm in society, enforced ultimately by sanctions of a social nature, rather than by the government. But again, until that stage is reached and maintained, policy makers will still want to know whether and to what extent the actions of voluntary agents in aggregate are having the desired effect.[7]

Political acceptability and feasibility

The severity of many environmental problems, especially those to which agriculture contributes, are heavily influenced by stochastic phenomena, such as the weather. Uniform standards and charges that are non-targeted – *i.e.*, not site- or situation-specific – are relatively inefficient instruments for dealing with such variability. On the other hand, any public law approach that involves frequent tinkering with constraints or penalties, besides requiring a considerable effort in order to communicate changes to those expected to comply with the law, in general risks reducing their acceptance of it. Or, as Rousseau put it almost two-and-a-half centuries ago: "... men soon come to despise laws which can be changed every day..." (Rousseau, 1984 [1755]).

More generally, delegating greater responsibility over environmental and resource policy to social institutions may prove to be the only pragmatic way for governments to deal with the number and complexity of environmental issues that they are being asked to address. As pointed out in a recent report by the Netherlands Scientific Council for Government Policy (WRR, 1992), the trend is for environmental policy to encompass increasing numbers of activities. Yet "to attempt to restrict all these activities using regulations governed by public law would eventually undermine one of the principles of our current social order

– freedom to act as long as it is not [legally] forbidden …". Rather, "Such restriction must be built into the social structure and order wherever possible" [p. 23; italics in the original]. A more recent study by the (United States) President's Council on Sustainable Development (1996) expresses much the same sentiment.

The conflicts over natural resources and their management are exceeding the capacity of institutions, processes, and existing mechanisms to solve them. Primary venues for addressing disputes tend to stress conflicts, polarising stakeholders and dividing communities. Litigatory approaches often result in solutions favouring the very few and fail to address concerns of the broader array of stakeholders. Too often lacking is a mechanism to bring the many and varied interests together in a forum designed to identify common goals, values and other areas of interest. The fundamental strength of co-operative approaches to resolving natural resource disputes is that they encourage the various stakeholders to identify with a particular place, environment, and resource, and to take responsibility for it.

Many advocates of co-operative approaches, looking at the process of environmental internalisation from a dynamic perspective, see them as particularly appropriate as transitional measures, especially in situations where demands for changes in environmental performance are being imposed for the first time. In such cases, the technology – because it has not heretofore been in demand – may be underdeveloped. Moreover, the information available to government regulators on available options and on costs to the target group will usually be poor. During this period of instability and change, both the target group and the government may perceive mutual advantage in non-confrontational approaches that encourage members of the target group to co-operate with government experts in seeking practical, low-cost solutions to environmental problems. Even if the process leads ultimately to formal regulation, the expectation is that such regulation will more likely be acceptable to both the target group and the government. On the other hand, both parties may discover that voluntary action or self-regulation may be sufficient.

FACTORS INFLUENCING THE SUITABILITY OF VOLUNTARY CO-OPERATIVE APPROACHES TO SUSTAINABLE AGRICULTURE

As suggested by Table 1, the policy instruments available to governments are not totally disassociated from the chosen institutional arrangement. Yet most research has examined only the suitability of different policy instruments for affecting behaviour, including voluntary agreements and social instruments, thereby possibly confusing the question of which tool to use with the question of who can best wield it. Members of private groups can decide among themselves to levy charges on each other, to set quotas, or to regulate behaviour in some other way – such measures do not have to be imposed from an external authority.

Indeed, a particular instrument that may be suitable according to economic and technical criteria may in some cases be more easily applied by groups on their members than by the government. Such self-taxation and self-regulation within groups is more prevalent than commonly supposed (see, *e.g.*, Chapter 3).

That said, the conceptual frameworks that have been developed to help in the selection of environmental policy instruments can also shed some light on the choice among different institutional arrangements. One such framework has been proposed by The Netherlands Scientific Council for Government Policy (1992), henceforth referred to by its Dutch abbreviation, WRR. Environmental policy problems are classified according to three categories of what the WRR calls "situation" characteristics:

- *Recognisability*: the ease with which activities can be measured or assessed; the recognisability of the effects and causal relationships between impacts and damage, and of the damage itself, and environmental risks and uncertainties.

- *Structure*: The number of sources of the problem in the target group; the number of persons or firms to which policy should apply; and the geographical and time scale of the impacts.

- *Resistance*: Abatement costs to the target group (whether they are high or low in proportion to revenues, and whether they are similar or vary widely among the group's members, both of which are also a function of technological options); the current financial situation of the target group; the balance of political power between the target group and the government; and the extent to which objectives of the target group and the government diverge.

Other considerations could also be added to the classification criteria, such as the relative influence of weather and other random factors, and the nature of the environmental damage (or benefit) function.[8] The intention here is merely to illustrate the variety generated by different possible combinations of factors.

Agriculture's effects on the environment tend to be difficult to measure, or the causal relationships between activities and effects are difficult to measure, or the relationships may be understood but the activities themselves are difficult to observe. There are, however, important exceptions to this generalisation. The number of farming enterprises that contribute to a particular problem depend on its scale. In respect of emissions of methane or carbon dioxide, the numbers of contributing farms are clearly large – over 100 000 in most OECD countries. For environmental problems that are primarily local, however, the numbers of decision makers can often be small – as few as 10 or 20. Dealt with at this level, problems that are difficult to observe in the aggregate become more obvious, thereby converting a problem that is complex or diffuse into one that is clear or distributional. The "localness" of a problem also has a bearing on the degree of resistance. It is difficult to generalise about the size of abatement costs relative to farmer revenues, since they vary widely according to the nature of the problem and the technological options. But again, the smaller the geographic scale, the narrower the cost differences, which would suggest less likelihood of friction among members of the affected population over an appropriate response.

Many of the same situation characteristics that bear on the selection of policy instruments can also be shown to have relevance for the feasibility of collective approaches. The *geographical distribution* of environmental effects from individual farmer behaviour in relation to abatement costs, for example, can be shown to have an important influence on the potential gains from joint action. Figure I depicts in highly stylised form the total discounted damage (or benefits) per hectare caused by farming activities carried out over the course of one year as a function of distance from the centre of a single farm property. For simplicity of exposition, the graph presents only one dimension in space – a ray extending from a single point – and relates damages (or benefits) to the diffusion characteristics of the effect. Clearly, the distributional pattern over a given area of an actual environmental effect will not be symmetrical in all directions; neither will the responsiveness of the affected area to degradation (or to improvement) be the same everywhere or proportional to dose.

◆ Figure I. **Stylised representation of farming activities that create externalities**[1]

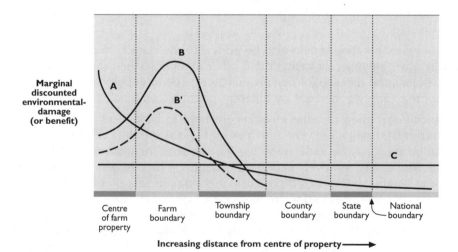

1. Assuming homogenous responses to environmental pressure.
Source: OECD Secretariat.

Three generic types of environmental problems are represented. The dispersion of chemical pesticides into the environment (Curve A), for example, is assumed to decline gradually with distance. Activities that affect wind-borne soil erosion (Curve B) in this example may result in relatively little damage to soils on the farm carrying out these activities, but considerable damage to neighbouring farms. The diffusion to the environment from emissions of gases that contribute to the greenhouse effect, such as methane or the soil fumigant methyl bromide, because it takes place within a global system, is widespread and therefore the marginal damage is shown as uniform over the globe (Curve C). Analogous examples could be given for the geographic distribution of positive externalities.

Just as the shape of the damage function for each of these environmental problems differ, so too will the costs of abatement faced by individual farmers. It is assumed here that the marginal cost of abatement is at least as high as the marginal damage being experienced at the farm level – that is, there are no net gains to be made for the farmer (call him Farmer Z) by modifying his farming practices even though such changes might reduce the damage occurring on his own property. (This, admittedly, may not always be a reasonable assumption, since Farmer Z may have better information on abatement costs than on damage costs, even those incurred on his own land, and therefore underestimate the latter.) Other land-owners living in the neighbourhood, by contrast, do stand to benefit from any changes in his farming practices that would lead to reduced environmental pressure on *their* properties. The likelihood that these neighbouring farmers would both have an incentive to, and be successful in, convincing Farmer Z to modify his farming practices in such a way depend on the characteristics of the externality concerned.

In considering, for example, an environmental damage function with a profile like that of Curve B, a large proportion of the environmental costs generated from Farmer Z's activities are externalised to his neighbours' farms. If he were the only farm creating such an externality, it might be worth the expense to affected land owners to pay him to take remedial measures (or, if the problem is serious enough, to take him to court). More typically he would himself be affected by similar externalities generated by these same neighbours, and those neighbours by other neighbours in turn. In theory, an elaborate system of transfer payments (or suits and counter-suits) could be arranged among all these farmers. Such a system would carry large transaction costs, however. More likely, they will devise some non-market and non-litigious solution that attempts to get every farmer in the area to commit to a common plan of action. The success of any such plan would hinge in part on the relatively high external costs compared with each farmer's costs of abatement.

On the other hand, in the case of an environmental damage function with a profile like that of Curve A, the net costs to Farmer Z of reducing his use of pesticides (in terms of lost revenues) could be quite high compared with the off-site impacts borne by surrounding land owners. However, the need for those pesticides may be affected as well by the practices used by neighbouring farmers.[9] Inappropriate pesticide use carried out by his neighbours could lead to increased pest resistance; insufficient pest control could increase his need for controls. In such a situation, farmers in the immediate area may have an incentive to work together to use their pesticides as judiciously as possible. Or, to take another example, growers in the area might each be willing to use a particular biological control agent, such as a parasitic insect, but only with assurances from their neighbours that they will not use a pesticide in such a way that it will harm the beneficial insect. In this type of "prisoners' dilemma" situation, efficient solutions may emerge through consultation and co-operation (Axelrod, 1984; Runge, 1984).

For diffusion/damage functions with profiles like that of Curve C, both the farmer and his neighbours will have little economic incentive to reduce their environmental impact to any large degree, since the share of local benefits to be gained, even collectively, are likely to be tiny compared with those of the rest of the world. Moreover, such benefits are unlikely to be realised by the current generation. Even ignoring this problem, the idea of all interested parties working out solutions within a group is made unfeasible by the large numbers involved – effectively, millions if not billions of people. Such transboundary issues invariably require co-ordination among larger, representative institutions – i.e., governments.

Figure 2 illustrates these same points in a slightly different way, by charting *cumulative* damage from an environmental pressure (here presumed to be a function only of its diffusion characteristics) as a function of encompassed area. Pressures of Type A and Type B are clearly predominantly local in their incidence; in the example, over half of the effects are confined to a relatively small area. Such problems lend themselves to local solutions. For one, in many cases the sources of the pressure can be observed by those affected by it. Second, the proportion of the benefits of any expenditures on improving the situation that spill over into outlying areas is likely to be modest. By contrast, with pressures of Type C, which are dispersed widely, local efforts will have little effect on local conditions.

Reality is of course more complicated than the simple model depicted in Figures 1 and 2. For one, farmers are faced with an array of damage functions at any one time, and have to set priorities among them. Individuals outside of an area may value a community's local environmental goods more than does the community itself. Introducing a temporal dimension into this simple model further complicates matters. Some externalities do not manifest themselves until many years following the activity that generated them, and the pathways may be obscure. Rising water tables and soil salinisation as a result of land clearing and

◆ Figure 2. **Cumulative damage or benefits of an environmental pressure as a function of dispersion over an area**[1]

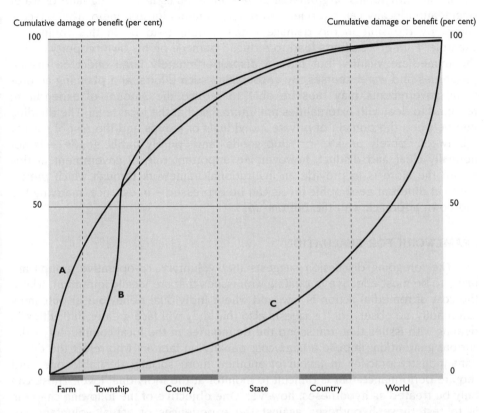

Cumulative damage or benefit (per cent)

Cumulative damage or benefit (per cent)

Increasing area from farm centre (log scale)

1. Assuming homogenous responses to environmental pressure.
Source: OECD Secretariat.

irrigation are perhaps good examples. Such externalities are difficult to internalise within a group, at least at the beginning stages, when the links between particular farm practices and their environmental consequences are not well understood. However, with improved information, small groups may still be able to address even these types of issues. It is not so much the time delay, but the uncertainty surrounding a long-term effect, that makes the search for efficient outcomes under such circumstances so difficult.

One further point can be made with reference to Figure 1. Efforts expended by private individuals or groups of private individuals in reducing land or water degradation locally can often lead to public benefits further afield. This effect is shown by the shift in the damage function from B to B'. In this conjectural example, Farmer Z's efforts lead to reduced damage on his own property and in the immediate vicinity, but have a disproportionately large effect on distant properties and water courses. By encouraging such efforts, and planning accordingly, governments may thus be able to reduce the amount of expenditure required to deal with externalities not internalised at the local level. The dividing line between the domain of private individuals or groups and the rest of society – between "purely private" or "club goods" and "purely public goods" – is not normally clear and distinct, however. An important role of government in this regard therefore is to provide an institutional framework through which preferences at different geographic levels can be expressed – in essence, marrying the top-down approach with the bottom-up.

FRAMEWORK FOR EVALUATION

The foregoing discussion suggests that voluntary, co-operative groups are likely to be most effective in dealing with issues that are locally important; where the cost of remedial action is low; and where individual behaviour or outcomes can readily be observed. It suggests also that they will face greater difficulties in dealing with issues that: transcend the boundaries of the local community or the current generation; impose a large cost penalty on farmers who reject the dominant industry practice in favour of another, more socially desirable one; and require performance that is difficult to monitor at the individual level. These can only be treated as hypotheses, however. One objective of the following chapters is to test these hypotheses against the experiences of actual voluntary, co-operative groups in action.

In order to be able to come up with policy relevant conclusions about these experiences, it is necessary to develop evaluation criteria for assessing both the performances of very disparate groups, and the efficiency and effectiveness of the policies that influence them. Because of the paucity of hard data on the groups briefly examined in this study, it is not feasible to develop and apply to them detailed criteria of a financial or environmental nature. However, much can be learned from looking at their structural features and the tasks that they undertake. Have they formed more or less spontaneously, or through the efforts of government? How many (active) members do they have? What were their reasons for choosing collective over individual action? What issues did they first attempt to tackle? How did they deal with these issues? How have the nature of these issues evolved over time – in particular, have they become more complex? It is necessary, as well, to be alert to the different structural features of the societies from which they have sprung, as such differences can have a major bearing on the social dynamics of the groups (formation of goals, trust in the leadership). These features in turn can influence the likely effectiveness of collective action.

The example given in Table 2 below, while pertaining to two contrasting groups in a non-OECD country, nonetheless is valuable as an example of how a pair of rural communities each deal with similar tasks requiring collective co-ordination and collective effort. Some attempt only to co-ordinate simple tasks, and are not highly effective at it; complex tasks are relegated to external (*i.e.*, government) agencies. Other groups are highly effective in mobilising community resources, with governments playing mainly a facilitating or supporting role.

Table 2. **Organisation of irrigation tasks in two villages in northwest India**[1]

Difficulty of task	Tasks performed by	
	Village 1	Village 2
Simple		
Observance of *warabandi* schedule (irrigation days)	Farmers, but leakages occur	Farmers
Channel maintenance	Farmers, but reluctance by purchaser households	Farmers
Liaison with Irrigation Department	Some individual farmers	None
Complex		
Water allocation	External agency	Farmers' association
Water distribution	External agency	Farmers' association
Conflict resolution	External agency, if at all	Farmers' association

1. Village 1 is characterised by a diverse social structure and a high rate of tenancy; Village 2 has the opposite characteristics.
Source: S. Sinha, "The conditions for collective action: land tenure and farmers' groups in the Rajasthan Canal Project", *Gatekeeper Series*, No. 57, International Institute for Environment and Development, University of Sussex, Brighton, UK, p. 12.

Regarding the efficiency and effectiveness of policy, a desirable set of evaluation criteria would include:

- *environmental effectiveness* – how much each instrument contributes to the achievement of the policy objective;

- *administrative and compliance costs and revenues* – the administrative and managerial cost burden imposed on both the administrative bodies responsible for applying the instrument and the economic agents subject to the instrument; revenues generated, and how they are used;
- *dynamic effects* – particularly whether it stimulates innovation;
- *"soft effects"* – whether, for example, the instrument changes attitudes and awareness, promotes capacity building, and aids the generation and diffusion of information;
- *economic efficiency* – the extent to which the instruments of the policy enable a more cost-effective achievement of policy objectives, compared with the alternatives;
- *wider economic effects* – impacts on the general price level, income distribution, employment and trade.

The extent to which these criteria can be applied depends on the period of observation and on the information available. Some criteria, such as wider economic effects, are entirely beyond the scope of the study. To treat that empirical question would require more data than can be assembled at this time. Similarly, effects on economic efficiency can only be considered from a qualitative perspective. And the data required to fully assess the environmental effects of the programmes considered here is still largely incomplete. However, information on programme costs, and details on the activities of the various groups, are beginning to emerge. Such information, if nothing else, permits at least a partial evaluation to be made of the effects and effectiveness of policies in the four country case studies that follow.

3

AUSTRALIA

BACKGROUND

Agriculture has been important to Australia since Europeans first settled there over 200 years ago. Today, agricultural commodities account for over 20 per cent of total merchandise exports. The climatic conditions under which farming takes place include humid tropical in the north and cool-maritime on the island of Tasmania. But the bulk of production of Australia's three leading agricultural commodities – beef, wheat and wool – takes place on the approximately 60 per cent of the country suitable for mixed farming and extensive grazing.

In the past, Australian governments focused assistance to agriculture through policies that attempted to mitigate the effects of unstable world markets and low world prices on farmers' returns. These policies included support for buffer stocks, price underwriting arrangements, tariffs, import embargoes and local-content schemes, and subsidies for inputs, such as fertiliser and water. Governments provided industry support in a way that led producers to expect government assistance, especially in times of drought or market downturn. This effectively sheltered farmers from risk and distorted market signals. Many farmers planned their operations on the basis of receiving such support and biased their production decisions towards products with higher level of assistance or reduced risk (DPIE, 1995a). Other support policies, such as tax breaks for clearing land and subsidies for irrigation, encouraged expansion of agriculture into fragile areas and exacerbated long-term changes in the soil hydrology.

Since the mid-1980s, the Federal Government has abolished market price support (except for milk) and tax-payer funded subsidies to fertiliser, and encouraged more reliance on market forces. Over the same time it has increased its funding of programmes for upgrading human capital in the farming sector, increasing social and economic opportunities for rural communities in general, and encouraged sustainable agricultural practices. Both the Federal Government and farmers' organisations in Australia believe that better environmental outcomes can be achieved by encouraging well-managed, commercially viable farms than by subsidising farming systems generally.

Box 1. **Dryland salinity in Australia**

Dryland salinity is a natural phenomenon in Australia, and is related ulti-mately to the ancient age of the continent's soils and a dry climate. Its incidence has increased markedly in recent years, however, mainly as a result of changes in land use which have led to increased percolation of water into groundwater recharge areas and consequently rising levels of saline water in discharge areas. As the water table rises, capillary action in the soil draws dissolved salts closer to the surface. When this water evaporates it may leave an encrustation of salt behind. When they reach the root zone, these salts can damage pastures and crops, induce changes in natural flora and fauna, reduce the service life of farm equipment and machinery, an increase the requirement for fertilisers. Over time, shrubs and trees die off, and erosion increases, leading to decreased soil fertility and increased off-farm effects, particularly siltation of rivers.

Dryland salinity began to emerge as a major problem in parts of Australia in the 1970s and 1980s – most notably in the wheat and sheep belt of Western Australia, and in parts of northern Victoria and southern New South Wales – and has been expanding ever since. In Western Australia the sudden surge in dryland salinity problems followed two to three decades after large areas of land began to be cleared for growing wheat, a development made possible by rapid advances in technology and government policies to encourage land clearing. A recent survey by the Land and Water Resources Research and Development Corporation (LWRRDC) estimates that some 1.2 million hectares were affected by dryland salinity problems in 1992-93. This represents just 0.3 per cent of Australia's total agricultural area, but a much larger per cent, perhaps one to two per cent, of its prime agricultural land. Even within these areas its incidence tends to be geo-graphically concentrated: some 16 per cent of broadacre farmers reported that dryland salinity was a significant problem on at least some parts of their proper-ties (Mues, Roper and Ockerby, 1994).

An important characteristic of dryland salinity is that impacts often occur far from the source of the problem. For example, the clearing of deep-rooted vegeta-tion in water recharge areas of a groundwater reservoir can lead to rising groundwater levels and seepage in discharge areas. The other complicating factor is that the time lag between cause and effect can be quite long – on the order of decades. This means that many farmers affected by problems of salinity can do little on their own to halt or reverse its progress. If the problem is not too severe, a farmer may be able to manage the problem through revegetating recharge areas, installing drainage systems or planting more salt-tolerant crops or grasses. Otherwise, efficient salinity management requires planning and co-ordinated action over a much larger area, such as an entire river catchment.

The LWRRDC expects that the area affected by dryland salinity in Australia will grow to between 2 million and 3 million hectares by the year 2010. The problem is expected to continue to worsen at least for the foreseeable future.

Although production-linked support has nearly disappeared in Australia, the effects of past policies, combined with farming systems inappropriate to Australian conditions and outbreaks of exotic pests (especially rabbits), have left a legacy of numerous land degradation problems (Gretton and Salma, 1996). Among the most locally serious of these is dryland salinity (see Box 1), but erosion, waterlogging, irrigation induced salinity and other forms of soil degradation are not uncommon. Loss of biological diversity is a feature of some rangeland areas. Typically, variation in these effects reflects interactions between climatic and socio-economic factors (Williams, Helyar, Greene and Hook, 1993).

GENESIS OF THE LANDCARE MOVEMENT

Farmer groups have been active in tackling land degradation since the drought years of the 1930s. These early groups were concerned primarily with soil erosion, their focus was narrow, and the direction of the programmes was determined largely by State Government agencies. The transition from government to community-driven farmer groups began in the 1980s. In a large number of cases, these groups were formed initially in response to land-related externalities, often created or exacerbated by farming activities, that were having a detrimental effect on farming, and generally getting worse: dryland salinity (and associated problems, such as water logging), animal pests, and erosion.

The State governments, nonetheless, were instrumental in encouraging and facilitating the formation of farmers' groups. The State of Victoria led the way in 1980 by establishing Farm Tree Groups to conserve and establish native vegetation. This initiative was closely followed by one in Western Australia. In 1982 Western Australia amended its *Soil Conservation Act* 1945 to encourage greater community involvement in land conservation by providing for the establishment of Land Conservation Districts (LDCs) and associated committees (LCDCs). As the number of farmer groups in these two States grew, so did community awareness of emerging environmental and land degradation issues. Participation in these groups remained voluntary and open to all, including members of local communities who were not farmers. This period also coincided with a growing and widespread acceptance that land degradation problems were serious and that many of these problems were outside the scope of any single farmer to solve.

In 1986 the Victorian Government established a new land protection scheme, under the registered name of *Land Care*. A main difference between Land Care and earlier programmes was that it was created to tackle a wide range of land protection issues, rather than single issues. By 1988, voluntary farmer groups had moved to the centre of land conservation efforts in Western Australia and Victoria. Similar groups had formed in other States, encouraged in some cases by the Commonwealth Government's National Soil Conservation Program, but these tended to be isolated examples. The term "landcare" was picked up by groups tackling local land degradation problems throughout Australia and was soon also used to describe the land conservation activities that they undertook. Nowadays it is used to describe a wide range of activities in the field of land and water management, as well as nature conservation (Australian Nature Conservation Agency *et al.*, 1995).

The idea for a truly national Landcare programme was first publicly mooted by the National Farmers' Federation (NFF) and the Australian Conservation Foundation (ACF) in early 1989 (see National Farmers' Federation, 1991). It received widespread support from all quarters. Subsequently, in his 1989 Statement on the Environment, the Prime Minister pronounced a "Decade of Landcare", which would commence in FY 1990. He also committed the Commonwealth Government to provide over A\$ 320 million (US\$250 million) for Landcare and related programmes over the ten years. In July 1990, the Commonwealth, States and Territories agreed to prepare plans setting out the roles, objectives and approaches that they, in co-operation with local communities, would take in implementing programmes to achieve sustainable land use for the Decade of Landcare. These plans were developed through consultation with individual landholders, community groups, non-government organisations, and local governments, and were progressively released in late 1991 and 1992. They, together with the National Overview, now comprise the *National Decade of Landcare Plan* (DPIE, 1995*b*, Appendix F, Section 3.1).

Since the National Decade of Landcare was proclaimed, the number and scope of community-based voluntary groups has expanded rapidly (Figure 3), as have the number of programmes created to support the plan. Over the whole of the country, some 34 per cent of farmers in the broadacre and dairy industries are involved with Landcare, and participation in some States is approaching 50 per cent (Chapman, 1997). In the rest of this chapter, only those groups and programmes with a primary focus on agriculture are discussed.

EXAMPLES OF LANDCARE GROUPS

There are now more than 4 250 landcare groups in Australia. A crucial characteristic of these groups is their voluntary nature, with membership open to all segments of the community. Groups elect their own officers and decide for themselves what issues to address and how to approach them, though specialists employed by the State governments are generally available to provide advice on techniques and funding. The following paragraphs briefly discuss several examples of Landcare groups, drawing from descriptions provided in Campbell (with Siepen, 1994) and in various reports and information fliers distributed by the Department of Primary Industries and Energy (Figure 4).

◆   Figure 3. **Growth in the number of landcare groups in Australia**

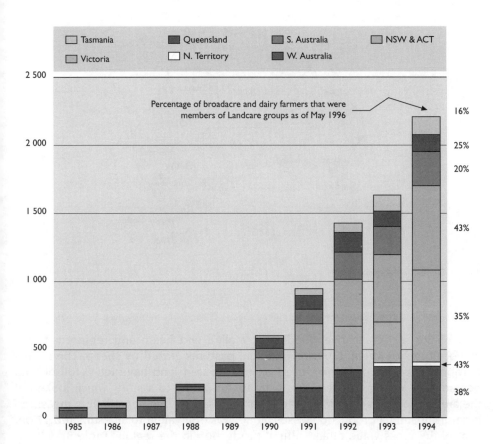

Source: Helen Alexander (1995), p. 12.

The Kalannie-Goodlands Conservation District Committee (LCDC)

The Kalannie-Goodlands Conservation District Committee (LCDC) incorporates seven sub-catchment groups (0.3 million hectares). It is located in the wheat belt of south-western Western Australia, an area roughly equal to England and Scotland combined. Clearing, mainly since the late 1940s, has removed 80 per

◆   Figure 4. **Location of selected landcare and related**
groups in Australia

Source: Campbell, with Siepen (1994).

cent of the region's native woodlands, mallee and heath and replaced it with monocultures of wheat, lupins and annual pastures grazed by sheep. The mainly European systems of farming used on this cleared land have left a toll of land degradation among the worst in Australia. According to a study which looked at the area's problems in 1988-89, the costs imposed by this degradation, measured in terms of annual production losses alone, was equivalent to an estimated 17 per cent of the gross value of agricultural production in the region (Table 3).

The Kalannie-Goodlands LCDC was formed in 1988 by a group of around 100 farmers who shared a collective concern about land degradation in the area. Already, about 10 per cent of the district was affected by obvious land degradation. The group was aware that a far greater area could become degraded unless farming systems were changed.

Some of the Kalannie-Goodlands LCDC's most innovative experiments have been in fund-raising. In order to raise money to help match an A$ 100 000 (US$78 000) grant from the Commonwealth government to support the preparation of a district plan, the group devised a "conservation rating system" for all landholders within the Conservation District linked to the unimproved capital value of their land. On the basis of this rating, landholders are charged on average A¢ 10.3 (US$0.08) per hectare per year. More recently, the LCDC has taken to

Table 3. **Land degradation in the Western Australian wheatbelt: situation as of 1988-89**

Degradation problem	% of land affected [1]	Annual production loses	
		A$ × 10^6	US$ × 10^6
Soil compaction	54	153	119
Water repellence	32	150	117
Salinity	3	105	82
Decline in soil structure	22	70	55
Waterlogging	11	90	70
Erosion and acidification	Small	47	37
Total	–	**615**	**480**

1. Cleared land.
Source: Western Australia Department of Agriculture (1991), as quoted in Campbell and Siepen (1994), p. 2.

issuing shares in capital assets purchased for particular projects. For example, it sold shares in a tree planting machine at A$ 100 each in order to match a A$ 7 500 (US$5 550) grant from the Western Australian State Landcare Program for its purchase. Shareholders are entitled to use the machine free-of-charge for four hours a year for four years.

Cloncurry Landcare Group

The shire of Cloncurry, in western Queensland, is home to one of the most isolated landcare groups in the country. The main agricultural activity is raising beef cattle on extensive pastures (farms range in size from 15 000 to 200 000 hectares); annual rainfall averages less than 500 millimetres. The Cloncurry Landcare Group was formed in 1989 following an initiative by the Shire Council, and the efforts of a core group of committed individuals. The group has about 150 members, few of whom attend business meetings regularly, but most of whom pursue landcare activities on their own properties. The group is led by an executive committee and operates through several subgroups. These subgroups initially considered a wide range of issues: stocking rates, control of pests (such as weeds, kangaroos, cane toads), land rehabilitation, education and awareness, and recycling. More recently they have narrowed their focus to just three key issues: stocking rates; total grazing pressure and land reclamation; and weeds.

Many of the group's activities involve either the creation of new knowledge (through workshops, market research, and mapping), or the sharing of equipment used for land rehabilitation. Its leaders consider that the group has had its greatest success dealing with the issue of weeds: mapping their location and educating land managers on ways to stop their spread. Campbell (with Siepen, 1994, p. 89) observes that "in the rangelands ... [the] real value [of landcare groups] is likely to be as a forum, a source of information, and support for finding ways of improving land management, rather than as a focus for co-operative work across property boundaries, as is more common ... where properties are much smaller and issues such as salinity demand a co-ordinated effort." He notes, however, that the young activists in the group are disappointed by the lack of broad community participation, which he attributes in part to the intrinsic difficulties in getting people across a large, sparsely populated area involved in collective decision-making.

Woady Yaloak Catchment Group

The Woady Yaloak River winds south from the Grampian Mountains west of Melbourne, Victoria, through 120 000 hectares of mixed farmland. Some 220 land holders who live and work in the river's catchment produce crops (mostly wheat), wool and livestock worth an estimated A$ 20 (US$16) million annually. Each year, however, the river carries 35 000 tonnes of salt and an unknown quantity of eroded soil into Lake Corangamite – Victoria's largest permanent inland lake and an important haven for wildlife. Already 900 hectares of farmland have been lost to rising salt, 80 kilometres of erosion gullies have been formed, and some 2 000 hectares suffer from sheet and tunnel erosion.

No less than six Landcare groups are active in the area, under the loose umbrella of the Woady Yaloak Catchment Group. The Group is more than half-way through a project to reduce the amount of salt and silt entering the river – the first such catchment project in Australia. Farmers in the catchment were heavily involved in developing the plan, which includes sowing pastures, planting trees and redoubling efforts to control rabbits and weeds. Of the A$ 446 000 (US$335 000) expected to be spent in connection with the project, A$ 145 000 will come from a private sponsor, the aluminium producer Alcoa Australia.

The Cundumbul Landcare Group

The Cundumbul Landcare Group was formed in 1991 by the ten farmers who comprised the membership of the local Bush Fire Brigade. The farmers got together initially to try to determine what was causing high salinity levels in a creek that ran through three of their properties. With the help of officers from the NSW State Government, the group started by drawing up property management plans for all ten farms. This initial process is credited with providing a secondary benefit: getting the farmers to learn more generally about problems on each others' land.

The Cundumbul group went on to organise Landcare speakers, workshops and field days. In 1992 the group began trials on all ten properties, looking at salinity, acidity and soil structure, and at ways to collect, propagate and plant trees and pasture plants that would help in arresting soil degradation. Unemployed youths funded under the Commonwealth Government's Landcare and Environment Action Program (LEAP) were brought in to assist in the planting. Since then, some 21 000 salt bush plants have been planted, along with several thousand trees.

Farm Management 500

Farm Management 500 (FM500) was established in 1991 to accelerate the adoption of effective farm business management practices. It is a commercially funded group extension programme involving over 500 farming families, 15 agricultural consultants and five agribusiness corporations from southern New South Wales, Victoria and South Australia. It operates through a process of 10 to 12 farming families working in groups, sharing information and learning from each other. The initiative grew out of an earlier project set up by two farm management consultants in 1986, called FarmFacts, which involved 80 farmers in a project to learn how to use computers for farm business management. At the end of this project, participants were able to develop business plans for their farms, along with farm monitoring and record-keeping systems, and to compare their own data against those of other farms.

The success of FarmFacts prompted the sponsors to expand the project to include a wider area and more farmers. FM500 got underway in earnest in 1992, with collaborative funding from several agribusiness, finance and insurance organisations, as well as from the Commonwealth Government, Charles Stuart University, and of course the farmers themselves. As Campbell (1994) points out, the benefits the private companies expected to reap from their involvement included using the FM500 groups to conduct product trials, and getting feedback from some of the region's most progressive farmers.

The aim of FM500 is "to increase the viability of both farms and farmers by harnessing the power of group learning among peer groups of farmers, facilitated by experienced consultants" (Campbell, 1994). In 1993 estate planning and retirement issues was the major focus. Out of this effort, the groups developed a set of benchmarks and business indicators to link farming systems to profit and sustainable development. The indicators include the following:

- net profit for the year/profit and loss; balance sheet, extent of borrowings, assets/liabilities/equity;
- comparison of budget with actual cash flow; operating surplus; cash in bank;

- yields from paddock/crops/stock/wool production;
- prices for inputs and products;
- ability to fund lifestyle; and
- the physical state of pastures/crops/property/sustainability.

More recently, the group has started up a project that it calls "FAST", which stands for "Farm Management 500 and Sustainable Technology". The FAST project attempts to link science with practice and farm profit with sustainability. The project involves collecting and analysing data on paddocks in a diverse range of soil types, climates and management practices from 60 properties in the south-eastern sheep and wheat zone. The study aims to evaluate the financial performance of each farming system and seeks answers to such questions as: which farming systems consistently perform in the top 25 per cent? do farmers with low disposable incomes borrow more or live off depreciation? can group members relate to financial benchmarks and sustainability indicators and use them for meaningful decision-making?

One important result that the FAST project is expected to yield is a set of computer packages on how to manage farms in such a way as to sustain profitable production without causing major damage to the environment. A comprehensive database and geographic information system will be set up that will allow farmers to assess the relative importance of, say, minimum tillage and crop sequence for different soil types or districts.

GOVERNMENT POLICY: THE NATIONAL LANDCARE PROGRAM (NLP)

Under the Australian Constitution, responsibility for the management of land and water resources rests primarily with the governments of the six States and two Territories. Agencies of these governments perform a variety of activities relating to land use, agricultural extension, environmental protection and the conservation of nature. Within this framework, local governments have responsibility for making many planning and management decisions within their jurisdictions. At the national level, the Commonwealth Departments of Primary Industries and Energy (DPIE) and of the Environment, Sport and Territories (DEST) are the main agencies responsible for environmental and resource management.

Most of the national level policies providing strategic support to community landcare groups are administered through the Commonwealth's National Landcare Program (NLP). The NLP was formed in 1992 through the amalgamation of several programmes, including the former National Soil Conservation Program (NSCP) and the former Federal Water Resources Assistance Program (FWRAP). The NLP has three components: a Community Landcare component; a Commonwealth/State/Territory component; and a national component. These are summarised below.

Community component

The Community component of the NLP is the one that provides the bulk of direct assistance to community landcare groups. It was established to assist these groups and local Government authorities to undertake activities for the sustainable management of land, water, vegetation resources, biological diversity and cultural heritage in their local area. It consists of the community elements of the land and water management activities administered by the Commonwealth Department of Primary Industries and Energy; various nature conservation programmes (such as Save the Bush, One Billion Trees and Waterwatch) administered by the Australian Nature Conservation Agency; and the Murray-Darling Basin Natural Resources Management Strategy (the "Integrated Catchment Management Program"), administered by the Murray-Darling Basin Commission. The programme aims to encourage landholders to identify the soil, water, vegetation management and nature conservation problems which concern them and to solve these problems effectively.

The Community component provides grants for group activities that contribute to problem identification and solution through activities such as: preparing farm plans; conducting demonstrations and trials; exchanging information and upgrading skills; and monitoring the conditions of natural resources. Community groups and local Government authorities can receive funding for management and conservation projects where one or more of the following conditions apply:

- the benefits can be shared by the local community;
- there is strong community support and contribution;
- the project will help develop relevant experience, knowledge and skills in the community;
- the project is part of a long-term natural resource management strategy, for example a catchment, regional, vegetation or other management plan;
- the results of the project contribute to national objectives of sustainable resource management; or
- the results will be communicated to other interested individuals and organisations.

Commonwealth/State/Territory Component

The Commonwealth/State/Territory component of the NLP provides support for those projects undertaken jointly by the Commonwealth and the States and Territories to achieve broad strategies or to co-ordinate activity within or between landcare groups or regions. It also supports projects that incorporate major investments in capital works. The key objective of the Commonwealth/State/Territory component of the NLP is to promote and stimulate change through partnerships that will result in improved natural resource management. The Commonwealth attaches particular importance to projects that:

- improve the ability of communities and resource managers to manage land, water and related vegetation in a sustainable and self-reliant manner;
- address the causes of environmental and resource degradation, rather than the symptoms;
- promote integrated approaches to catchment or regional planning and management;
- increase economic efficiency and cost effectiveness, including improvements in the performance of resource management agencies and public utilities;
- demonstrate innovative approaches to natural resource management;
- encourage the use of land, water and related vegetation resources within their capabilities; and
- improve the management of ground and surface water quality in key areas (particularly in sensitive ecological systems) such as through improved land management practices.

National Component

The National Component is the only component of the NLP funded directly by the Commonwealth (administered by the DPIE) rather than going through the State/Territory Assessment Panel (SAP) review process (see below). It funds projects that are primarily a Commonwealth responsibility. In particular it targets areas where additional effort is required to achieve national priorities. A project benefiting from financial assistance provided under the National Component must have a national focus, with benefits in more than one State, and should address issues of national significance. Investigation studies, or trials of innovative resource management policies or practices, can also receive funding.

The National Component is aimed at providing catalytic and facilitative funding to review, pilot or demonstrate innovative natural resource management policies or practices of significant national interest. It is not intended to provide ongoing administrative support nor does it fund research, although it allows scope for collaboration with research and development corporations.

FUNDING

Project applications are assessed by State or Territory Assessment Panels (SAPs) with representatives from organisations that have an interest in landcare, including community, farmer and conservation groups, indigenous peoples, local governments, non-governmental organisations, and State agencies responsible for vegetation, land and water management and for nature conservation. Final decisions on the recommendations of SAPs are made by the Commonwealth Minister for Primary Industries and Energy and the Minister for Environment, Sport and Territories.

Long-term support for resource management activities is not available through the NLP, and project funding is normally provided for a maximum of three years. The intention of this rule is to allow flexibility for NLP funds to be directed to new priorities as needs change. NLP funds are not intended to substitute for the resource management responsibilities of other levels of government or individual resource managers.

The bulk of the financial assistance provided by the Commonwealth government under the National Landcare Plan (NLP) is channelled through State agencies, with the rest given to programmes under the Community component (Table 4). Of the roughly A$ 19 million (US$14 million) provided under this component, slightly more than half goes to these groups directly. The rest is used mainly to help pay for support services (e.g., salaries of Landcare facilitators and technical advisors), the dissemination of information, and administrative costs. Table 5 shows the payments made in 1993/94 to three of the groups described in the examples cited above. The first five are typical of the approximately 500 community landcare projects that received Commonwealth assistance through the NLP's Community component in that year (the last project listed was funded through the National Component). Payments thus averaged around A$ 10 000 (US$7 500) per project; few exceeded A$ 50 000.

Most of the resources used in land and water resource improvement at the farm level are provided by the members of the landcare groups themselves, or by their local communities. Groups who apply for project funding under the Community component of the NLP must be willing to cover at least one-third of the total cost of the project, either in cash or in kind. Most groups contribute in kind, in the form of labour (time), use of equipment, materials, travel, etc. Where cash is involved, the beneficiaries are usually expected to provide the bulk of the contribution, with the remainder distributed more or less evenly among the rest of the group's membership. Some studies have shown that, besides the resources provided in connection with NLP projects, farmers are investing time and financial capital of their own. Donaldson (1995), for example, claims that the New South Wales Salt Action programme acted as a catalyst to private spending, showing a private:government spending ratio of up to 11:1. In only a small proportion of landcare groups is such private money raised collectively and on an on-going basis. The Kalannie-Goodlands land conservation levy is one such example.

In the future, some of the private and government funds to support Landcare will come from one pool: the National Heritage Trust. In June 1996 the Commonwealth government introduced legislation into Parliament to establish the Trust,

Table 4. **Commonwealth allocations to Landcare-related programmes, 1993/94-1996/97**

Millions of current Australian dollars

Programme element	1993/94	1994/95	1995/96	1996/97p
SUBTOTAL: COMMUNITY COMPONENT	**13.91**	**15.08**	**16.30**	**18.74**
Community Landcare Groups	4.39	5.46	7.80	10.47
Support for Landcare	6.36	6.34	5.72	6.08
Capacity Building and Participation	3.16	2.84	2.46	1.81
Landcare evaluation co-ordinators	–	0.44	0.32	0.38
SUBTOTAL: C'WEALTH-STATE COMPONENT	**49.82**	**49.79**	**56.60**	**38.29**
Catchment Management	**21.23**	**19.21**	**17.57**	**7.70**
Catchment Management and Planning	9.90	9.08	9.14	4.29
Floodplain Management	11.33	10.13	8.43	3.41
Land Management and Sustainable Agriculture	**11.97**	**14.89**	**16.35**	**16.81**
Land and water resource assessment	3.74	3.87	3.37	2.71
Property management planning	2.84	3.57	4.50	7.72
Other land management	5.39	7.45	8.48	6.38
Regional Initiatives	**6.20**	**7.71**	**7.54**	**9.98**
Highlands irrigation rehabilitation	1.80	1.80	2.20	1.80
Great Artesian Basin	1.20	0.90	0.90	0.83
Mt. Lofty (incl. Myponga)	3.20	–	1.15	1.04
SW Queensland Regional Program	–	0.89	0.79	0.94
Wimmera Mallee Pipeline Stage II	–	2.55	–	–
Other regional initiatives	–	1.57	2.50	5.37
Water Services	**10.43**	**7.99**	**15.15**	**3.81**
Darling Basin "Hot Spots"	1.62	2.26	7.38	0.34
Country Towns (water and wastewater)	8.50	4.28	6.70	3.47
Other	0.31	1.45	1.07	–
SUBTOTAL NATIONAL AND OTHER	**5.74**	**8.22**	**9.38**	**3.53**
TOTAL – ALLOCATIONS (SOIL AND WATER)	***69.47***	***73.09***	***82.28***	***60.56***

p = Preliminary.
Source: Programme estimates from the Commonwealth Department of Primary Industries and Energy, 1997.

"to provide for the conservation, sustainable use and repair of Australia's natural environment" (OECD, 1997*b*, p. 56). A$ 1.1 billion (US$860 million) from the partial sale of Telstra, the government-owned telecommunications company, will be placed in the Trust. It is expected that other stakeholders and beneficiaries will also contribute to the Trust. In all, the Trust is expected to provide some A$ 279 million (US$220 million) in additional funding to the NLP over the next five years.

Table 5. **Commonwealth allocations to selected Landcare-related projects, 1993/94**

Landcare Group	Project	Total Payments (A$)
Kalannie/Goodlands Land Conservation District	Remuneration for part-time co-ordinator	13 585
Kalannie and Goodlands regions	Land hazard evaluations and surveys	40 340
Kalannie and Goodlands regions	Resource inventory for integrated property and catchment plans	13 300
Cundumbul Landcare group	Cundumbul soil structure project	2 950
Cundumbul Landcare group	Resource inventory for integrated property and catchment plans	15 000
Farm Management 500	"FAST" (FM500 and sustainable technology) project	112 061

Source: Data source: Department of Primary Industries and Energy, *National Landcare Program – Report on the Operations of the Land and Water Elements*, 1993-94, Australian Government Publishing Service, 1995.

LINKS WITH OTHER PROGRAMMES

While the NLP is the main government programme supporting community landcare groups, other programmes are working with or through these groups to achieve objectives related to sustainable agriculture. These include in particular the Rural Adjustment Scheme (RAS), and various property management planning initiatives. Figure 5 shows how these programmes are related to each other, both in terms of the nature of their expected benefits and whether they achieve their goals through individual or collective action, or both. The following paragraphs look at how the NLP and property management are being integrated into rural adjustment and river basin planning.

Linkages between landcare and region-based structural adjustment policy

Since 1995, the Commonwealth Government has been trying to integrate its landcare efforts with structural adjustment, in an effort to link sustainable resource management to broader economic and social development. This means, in particular, combining resources available under the NLP with those under the Rural Adjustment Scheme (RAS). The RAS is the Commonwealth's main vehicle for encouraging structural adjustment in the farm sector. For many years, the RAS mainly assisted individuals. Recent reforms, however, allow it to support group-based approaches as well. One of the first applications of this new approach was the South West Queensland Initiative (SWQI), a project jointly funded by the Federal and Queensland Governments and landholders.[10] The region faces

◆ Figure 5. ***Intended coverage of natural resources management and structural adjustment programmes in Australia***

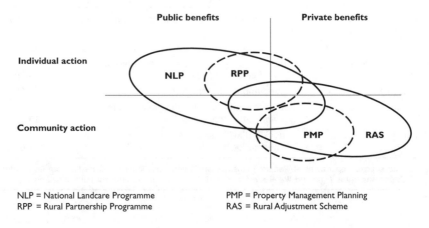

NLP = National Landcare Programme
RPP = Rural Partnership Programme

PMP = Property Management Planning
RAS = Rural Adjustment Scheme

Source: Land Management Task Force, Managing for the Future, Commonwealth Department of Primary Industries and Energy, Canberra, ACT, Australia, 1995.

severe economic, environmental and social problems arising from widespread land degradation, the unsustainable use of bore water from the Great Artesian Basin (the largest artesian groundwater reserve in the world), and poor economic prospects for a large number of pastoral enterprises.

Assistance is being provided through the NLP (A$ 2.8 million over three years) to help landholders better manage artesian water resources and to adopt improved property planning and land management practices. Improved management of stock watering systems, through the replacement of open drains with PVC pipe, has already enabled landholders to better control grazing pressure and to cut wastage by up to 95 per cent.[11] Funding is also being provided under the RAS to assist economically viable landholders to improve the productivity and sustainability of their land. Land holders who are adjudged not to be economically viable are offered assistance under the RAS to help leave the pastoral industry. This adjustment process is further underpinned by efforts to reform land leasehold arrangements. Other issues being dealt with under the SWQI include the sustainable harvesting of wildlife and the conservation of native flora and fauna.[12]

Linkages between property management planning and catchment and regional planning

Property management planning (PMP) is a core activity of many landcare groups in Australia, and is encouraged by numerous federal and state programmes, particularly the NLP. In recent years, bodies responsible for physical planning, especially at the level of rainwater catchments, have come to realise the advantages of integrating PMP, which is aimed at individual properties, into the broader process of catchment management planning. As PMP and catchment management planning use similar processes and approaches to achieve better natural resource outcomes, their integration would appear to offer opportunities for synergy. Like PMP, catchment management planning is relatively new, so examples of such integration are few. However, the report of the Commonwealth's Land Management Task Force (DPIE, 1995c) cites four examples of where the physical aspects of property and catchment management are now linked. This has tended to occur where neighbourhood groupings of landholders – often members of landcare groups – have undertaken PMP workshops in a catchment or sub-catchment context.

At a much larger scale, community groups are also becoming integrated into regional natural resource planning and management processes, the prime example of which is the Murray-Darling Basin Initiative. The Murray-Darling Basin drains a region some 1 450 kilometres long and up to 1 000 kilometres wide, or one-seventh of Australia's land mass. About 90 per cent of the nation's irrigated food crop is produced within the Basin. Water is mainly a State responsibility in Australia, but because the Murray-Darling Basin is so large – crossing four state boundaries and encompassing the Australian Capital Territory – its management clearly requires co-ordination at an inter-governmental level.

The Murray-Darling Basin Initiative, established in 1985 jointly by the Federal Government and the governments of New South Wales, Victoria and South Australia (Queensland joined in 1992), is an attempt to address this need. Responsibility for providing strategic policy direction for the Initiative is provided by a Ministerial Council, comprised of relevant Ministers from each participating government. Community groups get to express their views through a Community Advisory Committee (CAC), membership in which is drawn from local governments, industry organisations, and catchment groups (which usually include representatives from landcare groups). The Murray-Darling Basin Commission, which has responsibility for developing and implementing Council initiatives, also encourages and provides mechanisms for community involvement in natural resource management in the Basin. Priority is given to issues where co-operative action between Basin governments and the community is required, or where actions in one State may affect other parts of the basin. Such issues include irrigation, and the off-site effects of land management practices.[13]

In an experiment that may well be the first of its kind, in April 1996 the 1 800 farming enterprises and 25 000 residents of a 95 000 hectare region bordering the Murray River in southern NSW contracted with the State and local governments to tax themselves over the next 30 years in order to raise funds for a community environmental land and water management plan (OECD, 1997b). The plan was overwhelmingly endorsed by the local community, and involved more than 200 meetings with farmers. The bulk of the A$ 380 million (US$300 million) being raised to support projects under the plan will be provided by irrigation farmers, though municipalities and dryland farmers will also contribute to it. The plan addresses not only water supply, but also pollution from nutrients and pesticides. A new entity, Murray Irrigation Ltd., was created to be legally responsible for all activities related to expenditure, monitoring, and reporting, but will draw on representatives from all community sectors to oversee its activities. The state government's contribution to its implementation – A$ 60 million (US$47 million) over 15 years – is based on the proportion of public benefit (as opposed to private landholder benefit) that is expected to be achieved through the plan.

EVALUATIONS

At this point, more than half-way through the Decade of Landcare, a number of evaluations of the NLP have begun to emerge. The following highlights some of the initial findings.

Increasing awareness of land management issues

The goal of increasing awareness of issues holds central place in national policies to promote sustainable land use, and it is perhaps the most tangible accomplishment to date. In April 1993 two investigators (Curtis and De Lacy, 1996) surveyed all landholders in 12 sub-catchments in the north-east quadrant of the State of Victoria and found that respondents who were members of landcare groups exhibited significantly higher levels of awareness for almost all of the key land management issues listed than did non-members. Comparison of landcare and non-landcare participants revealed no significant differences in the holding of a stewardship or land ethic, however. Their results did tend to support claims that landcare group participation is making "an important contribution to enhancing the knowledge and skills of land managers through these activities as well as through contact between members, friends, relatives and neighbours" (p. 131).

Awareness of land management issues has also been increasing among the public at large. Between 1991 and 1993, for example, the percentage of Australians aware of Landcare almost doubled in rural areas (from 40 per cent to 70 per cent), and increased four-fold (from 10 per cent to 40 per cent) in urban areas (Landcare Australia Ltd., 1994). Landcare is also credited in Australia with widening the debate over nature conservation, from a concern primarily with heritage areas and natural parks to the proper management of the rural environment in all its forms. The increase in non-farmer community involvement in activities to promote sustainable land management can also be seen as an indirect effect of the landcare movement. According to the DPIE, an estimated 20 000 people, including students, land holders and community groups are now involved in regular monitoring of surface and groundwater resources.

Providing a new social focus for rural communities

In Government and independent assessments of landcare, some of the most often-cited benefits to rural communities relate to the social role it plays, particularly in isolated areas (see, e.g., Nelson and Mues, 1993). In the words of one National Landcare Facilitator:

> The importance of social cohesion in times of severe economic and environmental pressure and the need for community solidarity to accelerate change cannot be overemphasised. Too often sociology has been neglected in the development of public policy. The landcare group has been a vital social focus, in rural Australia in particular, but also in coastal and urban communities. ... Landcare has [become] a forum in which rural people have been able to do something in very rough times, and often taking the first step has meant achieving more than they had thought possible. As such, landcare groups are a powerful force in community development.

Changing institutional arrangements

According to one National Landcare Facilitator, the success of landcare should also be measured in terms of its impact on the institutional arrangements that will support sustainable development (Alexander, 1995). Research and development (R&D) appears to be a case in point. According to the Government's own evaluation, the way that R&D is undertaken in Australia has shifted more and more towards the sustainable management of natural resources. The mechanisms by which the results of R&D is communicated to landholders has also improved, in large part because they are themselves becoming more involved in R&D projects (DPIE, 1995b).

As described above, another important change has been in the role that communities play in total catchment management and in inter-State programmes, most notably those to improve the management of the Murray-Darling Basin. Whereas a decade ago policies were directed downwards, from national to regional to local, the trend over the last five years has been for policy formation to move increasingly in both directions. Communities are starting to work together across larger geographic units, such as regions or river-basin catchments, in planning, monitoring and implementing programmes.

Issues and activities

Particularly during the early years of the landcare movement, the majority of groups formed initially to address single issues, such as salinisation, pests and wind erosion. In addressing these issues, the groups tended to concentrate on farm planning (usually in a catchment-wide context) and on raising awareness of environmental issues in their communities. Case studies of successful landcare groups indicate that, as the they have developed and matured they have progressed to addressing the root causes, not just the symptoms, of the problems they encounter (Gorrie, 1995).

Changing land management practices

A long-term goal of the Decade of Landcare is for all public and private land users and land managers to understand the principles of sustainable land use and to apply them in their use and management decisions. Farm plans are seen as a necessary first step towards achieving such an understanding. According to a survey of landcare membership and practices, conducted by the Australian Bureau of Agriculture and Resource Economics (Mues, Roper and Ockerby, 1994), over half of farm operators in broadacre industries (*i.e.*, growers of wheat, beef and sheep) who were members of a landcare group possessed a farm plan in 1992-93, compared with only 22 per cent of broadacre farmers who were not members. The gap was even higher among dairy farmers: 50 per cent and 13 per cent, respectively. Overall, the share of broadacre and dairy farmers with farm plans was estimated to be around 30 per cent – a slight decline compared with the findings of a similar survey conducted in 1991-92 (Nelson and Mues, 1993). In terms of the information provided, however, the latest plans appear to be more complete. Over two-thirds contained information on the capability of the soils or land on the property; 45 per cent contained information on wildlife habitat or vegetation; and 40 per cent contained contingency plans for dealing with drought. Again, the farm plans of landcare group members were generally more comprehensive than those prepared by farmers who were not members of such groups.

Attitudes and awareness are not necessarily predictive of behaviour, however (Vanclay, 1992). To what extent are farmers following through with their plans? ABARE survey results suggest that the majority of both broadacre and dairy farmers consider their farm plans of either good value or some value in guiding decisions affecting land use. As well, those with farm plans appear to be implementing at least some of the changes to their properties called for in those plans (Figure 6). Members of landcare groups in both the broadacre and dairy industries were found to be more likely to be actively implementing their farm plans in 1992-93 than those who were not members (Mues, Roper and Ockerby, 1994). Planting trees was the most common activity.

◆   Figure 6. *Activities performed by Australian broadacre farmers in 1992-93
as part of the implementation of a farm plan*

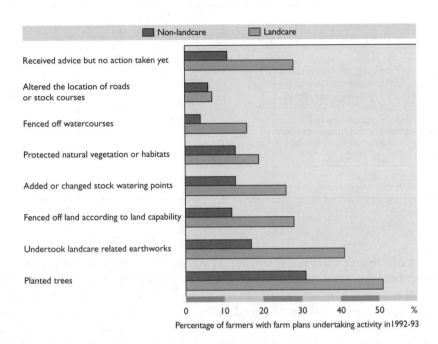

Data source: Mues, Roper and Ockerby (1994), p. 56.

The ABARE has also been conducting surveys since 1990-91 on different farm practices in order to track their adoption levels over time. Table 6 lists the results for a selection of these practices. Caution should be used in interpreting this table since farm practices that are appropriately part of a sustainable agricultural system will differ in importance from one area to another. In general it would appear that the rate of adoption of land conservation practices has been higher among farmers who are members of landcare groups than among those who are not. Such a finding was also confirmed by Curtis and De Lacy (1996) in their detailed survey of north-eastern Victoria. The exception in the ABARE study was the use of conservative stocking rates. However, the survey was based on farmers' perceptions, and no attempt was made to assess whether actual stocking rates were truly conservative in relation to the underlying carrying capacity of the property or of the prevailing seasonal conditions (Mues, Roper and Ockerby, 1994). It may even be the case that landcare members are more conservative than non-members in their estimation as to what stocking rates qualify as conservative.

Table 6. **Land conservation practices used by Australian broadacre farmers in 1990-91 and 1992-93**

Land conservation practice	1990-91 All farmers	1992-93	
		Non-landcare	Landcare
Used perennial pasture species	n.a.	63	70
Subdivided land into different classes	n.a.	33	51
Applied a conservative stocking rate	n.a.	84	76
Retained or incorporated crop stuble	41	37	46
Tested soil regularly	n.a.	28	40
Monitored water quality regularly	n.a.	14	36
Monitored pasture conditions regularly	n.a.	71	79
Managed crop rotation to minimise land degradation	n.a.	42	62
Planted trees and shrubs	11	38	64

n.a. = Data not available.
Sources: A. Nelson and C. Mues (1993); J. Ockerby and H. Roper (1994).

A more recent survey, carried out by ABARE in 1995-96, looked at the relationship between participation in training and adoption of selected farm management practices (Chapman, 1997). In all cases, the percentage of operators of Landcare farms who had participated in training to improve their farm management skills was higher than – often double – that of operators of non-Landcare farms. More than two-thirds had attended Landcare workshops and field days; other popular activities included participating in grower groups, enrolling in technical and further education courses, and attending property management planning seminars. The survey found also that adoption rates for sustainable farm management practices were strongly associated with participation in training. Farmers who had participated in three or more training activities over the past three years were often significantly more likely – in some cases twice as likely – to have adopted such farm management practices than farmers who had undergone no training. Indeed, for any given level of training, the differences in adoption rates between Landcare and non-Landcare operators were small. According to Chapman (1997), the results therefore suggest that decisions to adopt these practices is more closely associated with participation in training than with Landcare membership itself.

Of course, the adoption of "conservation measures" does not necessarily imply unambiguous consequences for the environment. Knopke and Harris (1991), for example, found that farmers adopting farming practices such as minimum tillage and direct drilling used significantly more herbicide, though also less fuel and machinery, than did farmers who were not using these practices – a finding confirmed subsequently by Nelson and Mues (1993). Whether the net results of this change of practice were positive or negative for the environment are difficult to assess: more careful management practices might in fact have led to reduced harm from herbicides, despite their increased use.

Changes in the environment

The ultimate test of the effectiveness of the Decade of Landcare will be the impact that it will have had on problems of land degradation and related water-quality impacts. At this stage, hard data on such impacts are not yet available.[14] What is available is information derived from farmers' perceptions. The 1992-93 ABARE survey, for example, asked farmers of their perceptions of the types of land degradation affecting their properties, and whether they saw these problems as increasing, stable or decreasing. Two problems of soil degradation – the spread of woody weeds and the decline in rangeland productivity – were seen by at least half of the farmers encountering these problems as increasing. Problems caused by wind erosion, by contrast, were perceived by the majority of surveyed farmers as either stable or decreasing in severity. Most other problems (such as water erosion, dryland salinity and soil acidification) were clearly perceived as either stable or still increasing (Mues, Roper and Ockerby, 1994).

A somewhat critical view is provided by Curtis and De Lacy (1996). Noting that, despite the positive influence of landcare, "a substantial proportion of Victorian landholders appear to be undertaking [only] limited amounts of land-care work", the authors conclude that "it would be difficult to sustain an argument that practices are being adopted at rates likely to produce major benefits at the landscape scale during the 10-year period of the Decade of Landcare Plan" (p. 135). In their view, arguments that increased funding to landcare would be a handout to land managers "ignore the community benefits of important landcare work such as revegetation, fencing water courses and establishing perennial grasses on steep hills ... fail to acknowledge that most land degradation problems have been inherited from previous generations, [and] deny the responsibility of government when government policies have contributed to many land degradation issues ..." (p. 136).

The question of who should bear the primary burden of paying for landcare is clearly one that is not likely to disappear. Australia's submission to the April 1995 meeting of the United Nation's Commission for Sustainable Development sets out clearly the Federal government's position on this issue: "The Commonwealth Government believes that, for rural landholdings and groups of such landholdings to be truly sustainable enterprises, they must be profitable enough to enable for the payment of on-farm works on a self-sustaining basis" (Commonwealth of Australia, 1995). Farmers, on the other hand, argue that it is unrealistic to expect the current generation of land managers to rectify all the mistakes of past managers and government policies. The Federal Government is therefore trying to encourage farmers to accept responsibility generally for the implementation of on-ground preventative measures, and believes that the mechanism to do this is through catchment-based and regional planning approaches, where both those who cause environmental damage and those who benefit from its remediation can be persuaded to pay their fair shares of the cost, along side a government contribution.

Other issues confronting policy makers are how to increase landcare participation above the current level (more than one-third of broad-acre farmers), how to encourage the adoption of sustainable land management practices both within and outside of landcare, and whether landcare is appropriate to all farmer groups or industries as a mechanism for information transfer and improving resource management.

4

CANADA

BACKGROUND

Canada is geographically the second largest country in the world. Most of its farming takes place within a 500-kilometre band north of its border with the United States. The country can be divided into four broad agricultural regions. In the western Prairies (Alberta, Manitoba and Saskatchewan), grain and beef cattle are the main products. In the Central (Ontario and Quebec) and Atlantic (New Brunswick, Nova Scotia, Prince Edward Island and Newfoundland) regions, the agriculture is more diverse, but dominated by intensive livestock production (including grain for feed), dairying, and horticulture. Agriculture in the western Province of British Columbia is regionally concentrated, with irrigated fruit production, dairying and cattle grazing predominant.

The range of environmental effects of agriculture in Canada reflects the country's great differences in soils, topography and settlement patterns. Owing to its cold winters, the need for insecticides is not as great as it is in some other OECD countries – only in Atlantic Canada, Ontario and Manitoba are insecticides used on more than 10 per cent of cultivated land – but over half of farms in Canada with cropland use herbicides (Dumanski *et al.*, 1994). Nutrient pollution from intensive livestock operations is locally important, particularly in the eastern provinces, and soil salinity afflicts pockets of land in various parts of the Prairie Provinces. In 1991 the inherent risk of *wind* erosion on bare soil was judged to be high to severe on 36 per cent of cultivated land in the Prairie Provinces. The risk of *water* erosion for Canada as a whole was high to severe on 20 per cent of the cultivated land; outside the Prairie Provinces more than 50 per cent of the cultivated land was rated high-to-severe risk. Implementation of soil conservation practices over the last decade has helped to reduce these inherent risks. The risk of wind erosion in the Prairies decreased by about 7 per cent, and the risk of water erosion nationally decreased by 11 per cent between 1981 and 1991. Wall *et al.* (1995) estimate that were all farms in 1991 using best management practices, less than 5 per cent of Canada's cultivated land would have been at high-to-severe risk from either wind or water erosion.

Canada's agricultural policies have undergone an important change in orientation in recent years. Agricultural support in Canada as a whole declined from a PSE of 46 per cent in 1990 to an estimated 22 per cent in 1996. The decline in support was the most dramatic for crops: the percentage PSE for crops (mainly wheat, coarse grains and oilseeds), which was 50 per cent in 1990, by 1996 had fallen to 16 per cent (OECD, 1997b). These trends in support reflect in part a reduction in market price support due largely to rising border prices, but also to reductions in payments under Canada's *Gross Revenue Insurance Plan* (GRIP) and in the transport subsidies provided under the *Western Grain Transportation* Act. Changes to policies relating to trade measures, some in response to the Free Trade Agreement between Canada and the United States (and later, to the North American Free Trade Agreement), have led to improved market access, putting further pressure on producers. Support for some livestock products remain relatively high, and for milk in particular it stood at 68 per cent in 1994. Most of this production is concentrated in the Provinces of Ontario and Quebec.

An important aspect of the general shift in Canada's agricultural policies away from production-linked assistance has been a greater focus on improving the sustainability of agriculture. The government's approach to sustainable agriculture encompasses not just agriculture but the entire agro-food industry. Its consultative process on policy development involves environmental groups, consumer groups, health-care professionals, producers, and governments at all levels. Under Canada's *Green Plan* (Government of Canada, 1990 and 1994) C$ 133 million (US$98 million) is to be spent between 1991/92 and 1996/97 on helping the sector make the transition to more environmentally sustainable practices. Of this amount, C$ 98 million (US$76 million) has been matched by equal contributions from the Provinces under joint agreements. Among other activities, the programme helps fund a Pest Management Alternatives Office, the costs of preparing environmental farm plans, watershed monitoring and support for provincial-level programmes to develop materials for promoting best environmental management practices.

GENESIS OF CANADA's RURAL CONSERVATION CLUBS

Farmers began forming rural conservation clubs in the Provinces of Ontario, Prince Edward Island, and Quebec in the early 1990s, encouraged by the federal-provincial funding under Canada's *Green Plan*. Analogous farmer-led organisations have also formed in the Prairie provinces and in British Columbia. The following describes in brief the main activities in Ontario, the Atlantic provinces, Quebec and British Columbia (Figure 7).

◆ Figure 7. **Locations of selected farmer-led initiatives in Canada**

Source: OECD Secretariat.

EXAMPLES OF FARMER-LED INITIATIVES

The Ontario Farm Environmental Coalition

In 1991 four groups – the Ontario Federation of Agriculture, the Christian Farmers' Federation of Ontario, AGCare (Agricultural Groups Concerned About Resources and the Environment), and the Ontario Farm Animal Council – decided to unite their activities in the area of sustainable farm management by forming The Ontario Farm Environmental Coalition (OFEC). The centrepiece of the OFEC's efforts to demonstrate Ontario farmers' strong commitment to their role as "stewards of the land" is their Environmental Farm Plan (EFP) programme.[15] The goal of the EFP programme is to help farmers develop practical plans for operating their farms in ways that are environmentally responsible. As set out in its report *Our Farm Environmental Agenda*, published in January 1992, the OFEC calls upon every farm family in the province (of which there are approximately 40 000) to complete an EFP by the year 2000 (Hilts, 1995).

To start off the process, farmers are expected to attend an introductory workshop at which they learn how to determine whether the soils on their farm can offset, or increase, potential risks to the environment. These workshops are intended to prepare farmers for the next step: filling out the their own EFP, using the 200-page workbook developed by the OFEC for that purpose. The workbook is based on 23 worksheets, each of which addresses a particular aspect of the farmstead (permanent structures, wells, storage facilities) or its land. In filling out these worksheets farmers are asked to give a rating to those conditions and practices that could affect the environment at different sites around their farms. Table 7 contains an extract of several of the questions and alternative ratings included in Worksheet #3 (pesticide storage and handling). The worksheets also indicate conditions that violate provincial legislation. Ratings marked fair or poor are expected to be addressed in the farmer's Action Plan, prepared later.

Farmers who have completed this step can then attend a second workshop, to consider alternative solutions to any potential problem areas they may have identified. Participation in such workshops is intended to prepare farmers for the fourth stage: the development of an Action Plan. As farmers work through these action plans, they determine what sorts of environmental problems can be expected to result from the combinations of natural conditions (*e.g.*, soil type and depth to groundwater table) and management practices specific to their farms. In so doing they are forced to think about what they need to do to solve or control these problems, and when to do it.

Farmers if they so choose can submit their plans to a Peer Review Committee, which will review the plan and make suggestions. These Peer Review Committees are composed mainly of local farmers. The plan is returned within a month to the farmer, who is then expected to put it into action and to re-evaluate it every year. As of early 1996, some 7 000 farmers in Ontario had attended the initial two-day workshops, of which around 5 000 had demonstrated to a Peer Review Committee that they had begun to implement their plans. The goal was for 12 000 farmers to attend the pair of workshops and to implement their EFPs by 1 March 1997.

The total cost to the Federal government of administering the five-year programme (including conducting workshops) is expected to be in the neighbourhood of C$ 10 million (US$7 million). The Provincial Government contributes equivalent amounts in kind, or in services. Initially, Agriculture Canada contributed C$ 500 (US$370) towards the cost of compensating farmers for carrying out actions identified in their personal Action Plans. (These funds are given to the

Table 7. **Sample from worksheet #3 of Ontario Environmental Farm Plan Workbook**

Condition or practice	Rating			
	Best (4)	Good (3)	Fair (2)	Poor (1)[1]
Total amount of pesticide stored	No pesticides stored at any time.	Less than 20 litres (or kg) of pesticide stored for longer than immediate use period.	20-200 litres (or kg) of pesticide stored for longer than immediate use period.	More than 200 litres (or kg) of pesticide stored for longer than immediate use period.
Distance from pesticide storage to nearest water source	Greater than 500 feet (150 metres).	200-500 feet (61-150 metres).	100-199 feet (31-60 metres).	Less than 100 feet (31 metres).
Distance from pesticide storage to well	Greater than 300 feet (92 metres).	76-300 feet (23-92 metres) if drilled well; 151-300 feet (46-92 metres) if bored or dug well.	50-75 feet (15-23 metres) if drilled well; 100-150 feet (31-46 metres) if bored or dug well.	Less than 50 feet (15 metres) if drilled well; less than 100 feet (31 metres) if bored or dug well.
Pesticide storage area	Stored in separate free-standing building or cabinet used only for pesticides.	Stored in a designated area, with separation walls, within a storage building.[2]	Stored in several designated areas, each with separation walls, within storage buildings.[2]	***Stored with human or animal food,*** or stored in residence.
Containment of leaks or spills in storage area	Impermeable floor (e.g., sealed concrete) that does not allow spills to soak into soil. Curb installed on floor to contain leaks and spills. No floor drain or floor drain to acceptable holding tank.	Impermeable floor with curb installed, but has cracks, allowing spills to get into soil. Or impermeable floor without cracks, but no curb installed. No floor drain or floor drain to acceptable holding tank.	Permeable (e.g., wooden) floor has cracks. Spills could contaminate wood or soil. No curbs installed. No floor drain or floor drain to acceptable holding tank.	Permeable surface (gravel or dirt floor); spills would contaminate soil. No curb installed. ***Has floor drain that leads to tile drain, surface water source, etc.***

1. Conditions indicated in bold italic type are in violation of Ontario's legislation.
2. Storage area may be adjacent to non-food items such as seed or farm equipment.
Source: Ontario Farm Environmental Coalition (1994).

OFEC, which in turn issues the actual incentive payment cheques.) Recently this figure was trebled to C$ 1 500 (US$1 100) in the hope that it would increase take-up rates. No figures are available on the total time and money expended by the farmers themselves in implementing their plans, but the amounts are believed to be equivalent to several thousands of dollars on some farms.

The Atlantic provinces

The Atlantic Farmers' Council (AFC) encompasses the four Atlantic provinces of New Brunswick, Newfoundland, Nova Scotia, and Prince Edward Island. In 1995 the AFC launched its own Environmental Farm Plan Initiative, with technical and administrative support provided by a federally funded agency, the Eastern Canadian Soil and Water Conservation Centre, and provincial departments of Agriculture and Environment in the four provinces. As in the Ontario Farm Environmental Coalition's Environmental Farm Plan (EFP) programme, the AFC's initiative is industry-led and participation is voluntary. Unlike that programme, no incentive payments are being used to encourage the completion of EFPs. Rather, consideration is being given to the creation of a special logo to identify the produce of farmers participating in the initiative. In addition, the AFC is establishing a regional database of information gathered (anonymously) from individual EFPs. It is expected that approximately 500 EFPs will have been completed by the end of 1996. The federal government has so far committed C$ 200 000 (US$145 000) to the project.

Quebec

Rural conservation groups in Quebec have also been involved in preparing environmental farm plans, though funding of their efforts has been more selective than in Ontario in that it has been targeted at those groups that appeared to be capable of achieving the best results. In general, the Provincial Government has been more directly involved in steering the environmental farm plan process, focusing on two watersheds where problems are most urgent. Federal funds have been used principally to hire professional advisors to assess the environmental damage caused by current agricultural practices and to prescribe a course of action for the watershed as a whole. The farm communities involved selected the advisors and were also closely involved in the development of the plans.

British Columbia

In British Columbia, farmers have established an Agricultural Environment Protection Council, which has two functions: to educate and assist producers in adopting codes of best management practice, and to monitor compliance with these codes. In effect, the Council operates both as a technical resource and as a self-policing body. The farmers that make up the Council are trained in a variety of pollution prevention techniques and understand the codes of practices and cost-effective means of achieving them. Should a producer not undertake the steps recommended by the Council and continue to pollute, peer pressure is applied in an attempt to get the producer to change his or her behaviour. Failing that, and as a last resort, the Council calls in officials from the Ministry of Environment to intervene.

EVALUATION

Given the small scale and recent history of farmer-led co-operative groups in Canada, there is little that can be said about them at this time. While several appear to have formed spontaneously, most were encouraged by the opportunities presented by the programmes flowing from the country's *Green Plan*. Their activities so far revolve around the promotion of environmental farm plans or codes of best management practice. In the case of Ontario's EFP programme, over 17 per cent of farm families have participated already in workshops since its inception, a share which is expected to grow over the next several years. The cost to taxpayers so far has been modest. The main benefits of the programme are likely to relate to changed attitudes and awareness and the generation and diffusion of information. Some of this information may help in the design and choice of indicators.

5

THE NETHERLANDS

BACKGROUND

Despite its small size, The Netherlands ranks consistently among the top four exporters of agricultural produce in the world. The sector is dominated by the production of livestock and dairy products (in total 56 per cent of gross value) and horticultural products (35 per cent); arable production accounts for a mere 8 per cent of output. The creation of a common market for agricultural produce within the European Community during the 1960s opened up new export opportunities, notably for growers of fresh vegetables, fruits and flowers; The Netherlands' central location and logistical infrastructure made it the focal point for European trade in these products. EC intervention in agricultural markets under the Common Agricultural Policy (CAP), especially for grains, sugar beets, milk, beef and sheepmeat, however, provided new stimuli to production, mainly in the form of price-support instruments. Such support, measured for the 12 countries then members of the European Community, varied between 45 per cent and 50 per cent during the 1986-1995 period (OECD, 1997b).

In order to extract high yields from a small amount of land, The Netherlands' 100 000 farms on the whole rely on intensive production methods, particularly those that make use of capital and chemical inputs. Indeed, many farming activities in The Netherlands are no longer dependent on the fertility of the land. Cut flowers, vegetables for fresh sale, and mushrooms are grown predominantly in glasshouses or other enclosed structures, largely on artificial substrate. The raising of pigs and poultry is also largely an indoor activity, and almost entirely dependent on purchased feed, much of it imported.

The environmental pressures created by these agricultural activities are extremely high. According to the OECD's latest *Environmental Performance Review* of The Netherlands (OECD, 1995), agriculture is now the dominant contributing sector to:

- *dispersion of toxic and hazardous pollutants*: 88 per cent of the total pressure from domestic sources, mainly due to chemical pesticides;
- *eutrophication of water and soil*: 84 per cent of the total pressure from domestic sources, mainly due to the spreading of manure and inorganic fertilisers;

- *acidification*: 61 per cent of the domestic pressure, in particular 91 per cent of ammonia emissions.

In responding to these pressures, the central Government has employed a wide range of measures. In order to reduce pesticide use in agriculture, for example, it has increased the amount of time between treatments for soil fumigants in arable farming from three years to four years; sponsored research into alternatives to current chemicals; and provided subsidies for conversion to "organic" methods of agriculture.

Laws to deal with eutrophication and acidification – both to a large extent related to the surplus manure problem – have been in place since the mid-1980s. Legislation was first introduced in 1984 aimed at halting the expansion of intensive livestock rearing operations, and so manure production, by limiting the number of animals and buildings on farms. But because farmers were able to anticipate this law by building up their breeding stocks to record levels and securing planning approval for expansions before the laws took effect, the Government was forced to modify its approach, and in 1986 it decided to limit the production of manure instead by restricting the amount of phosphate that could be applied to the ground. In addition, farmers were obliged to keep a detailed record of movements of organic fertilisers on and off their farms.

The governments' latest plan is to develop a system of minerals accounting, first as a management tool, and eventually as a means for assessing a levy on surplus nutrient production (OECD, 1995). Each year farmers would send in a provisional assessment of the total surplus of phosphorus and nitrogen produced on their farms, taking into account nutrients imported as feed and exported as product. From this surplus would then be subtracted estimated losses of nutrients to the environment that are deemed acceptable (such as estimated losses of nitrogen as N_2). A levy would be charged on the remaining, non-acceptable losses. The Government be applying the levy system to intensive livestock producers in 1998, and plans to extend it to other livestock producers in 2000. It is envisaged that the system would eventually be extended to all remaining farming activities. Dutch farmers, along with those in other member countries of the European Union, must also meet the requirements of the EU's Nitrate Directive (91/676), which sets out targets for the level of application of manure in the EU for zones that have been identified to be vulnerable to pollution from nitrates.

These measures, by forcing reductions in the amount of manure spread on arable land, are also helping to alleviate emissions of ammonia caused from the spreading of animal slurry. Since ammonia emissions are also sensitive to the timing of applications, however, the Government introduced a law in 1989 that forbids farmers from applying liquid manure on the surface of their fields during winter months. Rather, any slurry applied to the ground during this time has to be injected directly into the soil. Subsidies have been made available to help cover a portion of the cost to farmers of purchasing the necessary equipment.

These regulations and programmes have not been uncontroversial. Many environmental interest groups see them as too weak. Farmers, while accepting the need for greater attention to environmental problems caused by agriculture, regard them as too costly and too inflexible. They are particularly concerned about the laws regulating the storage, use and disposal of manure. They are, in short, looking for an alternative to the central-government directed approach to managing the environment.

GENESIS OF FARMER-LED ENVIRONMENTAL GROUPS IN THE NETHERLANDS

In 1990 the director of the National Council of Co-operatives (the NCR) proposed one such alternative approach. His idea was to encourage the formation of farmer-led associations or co-operatives that would take the initiative to integrate protection of the environment, including nature and landscape, as an essential component of the production process, and to take joint responsibility for the outcomes. The goals of these co-operatives would include not only improvement of the environment, but also improvement of the financial position of, and development opportunities available to, its members (van Dijk, 1990).

The idea was enthusiastically received by farmers across the country, and the first groups began to emerge in 1992. By early 1994, at least 30 groups had come into existence, with a membership of almost 2 000. Over the next two years their numbers grew rapidly, and as of October 1995 an estimated 60 declared co-operatives were in operation, with a membership of close to four thousand. These groups are located throughout the country and differ greatly in focus, size and structure. One thing they all have in common is a desire to apply locally-tailored solutions to national and regional environmental problems.

The emergence of farmer-led initiatives can be seen, to some extent, as a logical outgrowth of the sector's historical propensity to organise and regulate itself. Co-operatives for processing and marketing produce, for example, have been around for more than a century in some cases, and still account for the bulk of sector sales (OECD, 1994). Dutch farmers are also horizontally and vertically linked within a much broader, over-arching institutional structure, comprising commodity boards, industry boards, and farmers' unions (organised under the Agricultural Board or *Landbouwschap*). And they have a long-standing tradition of learning new techniques together and of sharing information, such as in the horticultural growers' "study groups" (OECD, 1994). Direct state intervention in the affairs of these organisations has customarily been minimal; the Government's view is that decisions should be left to organisations of entrepreneurs and employees where such organisations would be equally well or better suited than the Government in furthering their own interests (Boonekamp, 1992).

GOVERNMENT POLICY IN RESPECT OF THE FARMERS' ENVIRONMENTAL GROUPS

Central government policy towards farmers' environmental groups has generally been positive and supportive. Since 1993, the government's official position is that it wants to give more responsibility for applying locally tailored solutions to environmental protection to farmers. This view was also underscored by the Ministry of Agriculture, Nature Management and Fisheries (MLNV) in its Structural Scheme (*Groene Ruimte*), which stated that the roll of farmers as stewards of the countryside must be strengthened. The concept of devolving responsibility over countryside stewardship is also consonant with the government's more general inclination towards decentralisation.

In keeping with this spirit, in November 1994 the MLNV formally invited five farmer environmental groups[16] from around the country to submit proposals for co-operative projects addressed to environmental problems in their areas, including the preservation of nature and landscape. In preparing their plans the five groups received assistance from the Sociology Faculty of the national Agricultural University at Wageningen. On 15 March 1995 the five "action plans" (*plannen van aanpak*) were submitted to the MLNV. Each plan sets out one or more proposal for alternative approaches to managing their local environment – ranging from farm-level algal ponds to small-scale composting of plant residues – along with specific requests for project grants.

The five groups also asked for something else: special, albeit temporary, dispensations from environmental regulations, especially design standards pertaining to the handling and disposal of animal manure (Table 5). The argument put forward by the groups for these dispensations was basically that nation-wide regulations were too often inappropriate for their particular circumstances, and that in some cases they could even lead to greater damage to the natural environment or to the landscape. In other cases the restrictions, if strictly applied, would hinder the development of new techniques that were not envisaged when the rules were first drawn up. While committing themselves to the broad objectives of national policy, the groups asked for greater flexibility in meeting them.

According to the farmers' journal *Boerderij* (Dokter, 1995), the groups' requests for special status raised a number of difficult issues. First, the legal standing of the groups was unclear, particularly over the question of whether exemptions that a group obtains could be claimed by each farmer. Second, policymakers were concerned about whether the groups possessed the organisational strengths, internal cohesion and leadership authority necessary to compel members to comply with the agreed conditions. Finally, there was the question of how to handle proposals from other groups who might apply for special dispensations in the future. These could grow quickly to the hundreds by some estimates.

On 15 February 1996 the Minister for Agriculture, Nature Management and Fisheries gave the green light to the five eco-cooperatives' action plans, describing his decision as an "administrative experiment". On this basis, the five eco-cooperatives were granted the opportunity to demonstrate how current environmental goals could be achieved in a way that best accords with local conditions. As part of the deal, the MNLV made available money, legal advice and services to the groups. The Ministry also agreed to grant the eco-cooperatives release from certain regulations for which it was primarily responsible. For regulations that were the prime responsibility of other government agencies, the MNLV is to act as an intermediary.

Although the Minister supported the majority of the eco-coops' plans, a few sticking points remain to be worked out. Furthermore, given that each project is of a highly specific nature, it was acknowledged that individual agreements had to be reached regarding monitoring, inspection, and reviews. Interim reports on the experiment were issued in 1996 and 1997. These reports were positive about the results achieved to date, but concluded that it was still too early for the Government to be able to take a longer-term view regarding such approaches, or to consider proposals from other groups.

The following section provides some background on the five eco-cooperatives selected for the Government's "administrative experiment", and summarises their general and specific approaches to countryside stewardship.

THE ECO-COOPS AND THEIR PROPOSALS[17]

Eastermar's Landau Association (Vereniging Eastermar's Lânsdouwe) and the Achtkarspelen Agricultural Nature and Landscape Management Association (Vereniging Agrarish Natuur en Landschapsbeheer Achtkarspelen)

The Eastermar's Landau Association (VEL) was the first environmental cooperative established in The Netherlands along the lines proposed by van Dijk (see map, Figure 8). It covers a 1 900-hectare area in the north of the Province of Friesland, a pasture landscape criss-crossed by small canals and *houtwallen* (low, tree-covered earthen dikes, which also serve to corral livestock). These *houtwallen*, besides marking the boundaries between farm plots, have traditionally served as windbreaks, sources of firewood, habitat for game and other useful wildlife species. The main agricultural activity in the area is dairying. The farms, though small, are highly productive, with milk yields exceeding the Provincial and national averages by a considerable margin. Nine out of ten farms have a designated successor.

The VEL was formed officially on 27 March 1992 by a group of 56 farmers and other land-owners. Their common interest was in preserving the region's way of life and local landscape by coupling stewardship of the land to economically

◆ Figure 8. **Location of the five farmer environmental groups selected for special status in the Netherlands**

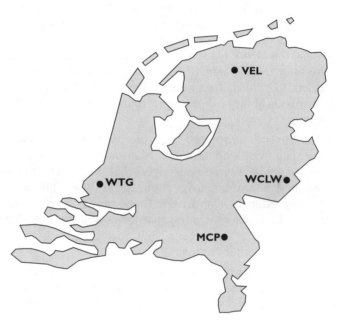

Source: OECD Secretariat.

sustainable agriculture. They also felt that current environmental regulations were not flexible enough to allow farmers to look for integrated and low-cost solutions to the area's environmental problems.

The group's first activity was to work out a more detailed blueprint for the area in the form of a model plan. To assist them in their efforts to prepare a model plan the central government provided them, along with each of the other five selected co-operatives, with a subsidy of Gld 50 000 (US$35 000). They also received help during the conceptual phase from the Wageningen Agricultural University and the Province of Friesland. This plan calls for an integrated approach to agriculture, nature, landscape and recreation; recognises the benefits of co-operation among farmers and other land-owners in this respect; and leaves room for individual farmers to take divergent approaches. A model farm-level plan (called a BOP) has also been drawn up. It assembles: a land-use chart; a list of possibilities for integrating and expanding activities in the short and long term;

the approach to sustainable land management to be taken; subsidy sources; and a financial balance sheet. The intention is that each farmer would develop a plan specific to his or her circumstances.

A major focus of the group is the maintenance of the region's deteriorating *houtwallen* and canals. Much of this work had previously been undertaken by farmers individually, but most of them are happy to see it organised collectively by the VEL: the farmers feel that such a collective approach is more efficient and is likely to attract a higher level of participation. The VEL is also co-operating with an agency connected with the Ministry of Agriculture, *Bureau Beeheer Landbouw-gronden*, and a nearby farmers' environmental group (VANLA), which has a membership of 90, on experiments to develop the natural value of the *houtwallen* in general and to improve their function as a habitat for wildlife and birds. A wildlife management plan has been developed for the area.

The VEL's membership look to the association as a gateway to policy makers. Its leadership thus plays an important co-ordinating role between the farmers and the different levels of government. It helps channel requests for project subsidies, obtains training and extension services, serves as a contact point for researchers, and mediates disputes between members. In the future, the members of the group see a role for the VEL in, among other activities, stimulating agri-tourism and organising a course in nature management.

In responding to the Government's invitation to submit action plan proposals, the VEL and the VANLA co-operated on a joint proposal, geared towards three areas: manure and ammonia emissions; nature management and protection; recreation and integrated water management. Specifically, they asked:

1. to be allowed to spray manure slurry on their fields during the whole month of September (currently the law requires stopping this practice by 1 September), and to use experimental low-emission slurry spraying equipment during the months of February through April (or through September if an additive is mixed into the slurry);

2. and for financial assistance for the collective repair and upkeep of *eizensingels* including the introduction of an experimental "quality premium".

The Government's response to these requests has been generally favourable. It approved the spraying of slurry during the latter half of September, although it saw no agricultural need to extend the spraying period beyond that. It also authorised Gld 200 000 (US$118 000) to help in the development of the groups' idea for a new spraying technique; if the results of the laboratory tests prove positive, practical trials would then be allowed. The government also allocated Gld 550 000 (US$325 000) to help pay for the deferred maintenance of *eizensingels*, and supported the idea of adjusting payments for the upkeep of landscape features to the quality of the results achieved.

Environment Co-operative "de Peel" (Milieucoöperatie de Peel)

The *Peelgebied* is an area of the Province of Noord-Brabant valued by the government for its natural and cultural landscape features. It is also an area that has seen a rapid growth in agriculture over the last several decades. Intensive livestock production, dairy farming, and glasshouse horticulture predominate. Animal husbandry in particular has exacted a heavy toll on the environment: the *Peelgebied* is one of the areas of The Netherlands most affected by eutrophication and acidification. By the beginning of the 1990s, nitrogen production by animals (cattle, sheep and pigs) for the Province of Noord-Brabant as a whole was around 400 kg per hectare – among the highest levels in Europe.

In 1992 the farmers and growers in the *Peelgebied* were being increasingly challenged by environmental groups – above all, the Working Group for the Protection of the Peel – to reduce their pressure on the environment. The general view of these environmental groups was that nature and farming were incompatible. In order to break this impasse the leadership of the two main farmers' organisations in the area established their own body, the Peel Working Party, the aim of which was "the preservation of economically viable agriculture in a valuable natural landscape". The Working Party's first action was to enter into discussions with the Working Group for the Protection of the Peel. It also commissioned a study of the environmental and agricultural problems in the area, including problems that farmers were experiencing with different environmental policy restrictions.

One of the conclusions of this study was that the current, undesirable situation was not constructive. People wanted a clear and unambiguous policy that created stimulating work and just conditions. Policy support in the form of financial assistance was seen as indispensable in this respect. Another conclusion was that a stronger farmers' organisation was needed. Thus on 2 July 1993 the Environment Co-operative "de Peel" (MCP) was officially established.

The co-operative's membership is drawn from the two main farmers' organisations in the area: the North Brabant Christian Farmers' Union (NCB) and the Southern Agricultural Society (ZLM). All members of these two organisations – about a thousand farmers – are nominally members of the MCP, though their participation in the co-operative's activities is not obligatory. The main activities of the MCP are carried out through working groups, each of which deals specifically with the problems faced by the three main industries. Overseeing these working groups is a steering group comprised of elected representatives from the NCB and the ZLM (Figure 9). An important function of the steering group is to review projects; those that are not approved by that body do not move on to the implementation phase.

◆ Figure 9. **Organisational structure of the Dutch environment co-operative "de Peel"**

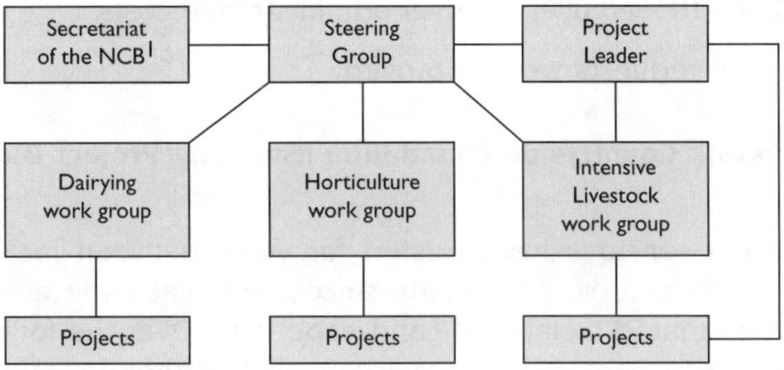

```
┌─────────────┐     ┌─────────────┐     ┌─────────────┐
│ Secretariat │─────│  Steering   │─────│   Project   │
│ of the NCB¹ │     │    Group    │     │   Leader    │
└─────────────┘     └─────────────┘     └─────────────┘
                                                        
┌─────────────┐     ┌─────────────┐     ┌─────────────┐
│  Dairying   │     │ Horticulture│     │  Intensive  │
│ work group  │     │ work group  │     │  Livestock  │
│             │     │             │     │ work group  │
└─────────────┘     └─────────────┘     └─────────────┘
                                                        
┌─────────────┐     ┌─────────────┐     ┌─────────────┐
│  Projects   │     │  Projects   │     │  Projects   │
└─────────────┘     └─────────────┘     └─────────────┘
```

1. Nordbrabanste Christelijke Boerenbond (North Brabant Christian Farmers' Union).
Source: Milieucoöperatie de Peel (1995).

One of the MCP's experiments involves setting up an algal pond on a farm belonging to one of its members. The algae growing in the pond are able to remove a large amount of the phosphorus and nitrogen contained in the slurry waste produced by the farm's piggery, yielding a residual stream of wastewater that the MCP claims is clean enough to use for stockwatering or irrigation, and a green biomass that can be dried and used as an organic fertiliser.

The MCP's action plan includes a wide-ranging set of projects. From the Government it asked for:

1. permission to apply actual emission factors (rather than theoretical) factors in the application of the government's ammonia law;

2. permission to work with the local Provincial Governments on a plan for implementing the "Nature Management Law" that would not necessarily conform with the permits required under that law;

3. help in introducing a detailed mineral accounting system (including a bonus/penalty system);

4. help in enabling manure to be processed at the individual farm level (as opposed to at a more aggregated level) – this would mean exemption from several technical norms regarding such things as the maximum allowable liquid fraction of manure spread on or worked into the ground;

5. financial support to help cover the costs of various projects proposed in the Peel Land Program (Gld 5.2 million, or US$3.5 million), and another Gld 90 000 (US$60 000) to cover administrative costs.

Most of these requests were approved.

Project Winterswijk Countryside Foundation (Stichting Project Buitengebeid Winterswijk)

In Winterswijk tensions have existed for years between local farmers and nature preservation groups, particularly since 1983, when the area surrounding the town was designated a National Landscape Park. Over the following decade, however, both sides came gradually to accept that neither agriculture nor nature preservation can benefit from polarisation. Farmers more and more saw the value in protecting the environment; environmental groups recognised that some agricultural activities were necessary to maintain what many of their members considered to be a "cultural" landscape. This new willingness to seek common ground was formalised in February 1993 with the creation of the Winterswijk Countryside Foundation (WCL).

The initiative for the WCL came largely from the leaders of the local farm organisations, who were worried about the future of farming in the area. The foundation's central role is to work with the various levels of government with jurisdiction over the area, in an effort to reach an acceptable solution regarding a number of aspects of existing legislation that the farmers consider to be unworkable. The WCL's action plan addresses concerns similar to those put forward in the plans of the VEL and VANLA. Specifically, they asked for:

1. the integration of the Government's ammonia reduction plan into farm-level environmental plans in a way appropriate to the region;

2. the establishment of tradable emission and deposit rights for ammonia and phosphate (in the form of a re-organisation and development fund);

3. a more flexible use of the Government's policy on nature management agreements, so as to allow their application at an area-wide, rather than a farm-specific level; a property tax exemption on areas maintained for nature; a more co-ordinated process for concluding management agreements; and the adaptation of current land-use planning procedures on an experimental basis;

4. a harmonisation and consolidation of instruments of financial support (*e.g.*, for enterprise relocation);

5. legal protection from restrictions to control farming odours, which might otherwise be imposed in the future as new, non-agricultural residences move into the area.

The Government approved most of these requests in one way or another, and undertook to look into the possibilities for working out ways to apply nature management agreements more flexibly. The request to change the tax status of land covered under nature management agreements was rejected as incompatible with the existing tax law.

Working Group Cultivation in the Ground (Werkgroep Telen in de Grond)

Unlike the other groups so far mentioned, the Working Group Cultivation in the Ground (WTG) is focussing on a particular sub-sector (glasshouse horticulture) rather than a particular region. The Group was formed out of the Netherlands' Foundation of Horticultural Study Groups (NTS), an organisation formed many years ago to promote the exchange of new knowledge among the horticultural community. The NTS regularly organises hundreds of lectures and excursions each year on topics of interest to growers (Blum, 1991). Many of the participants in these groups are also communicate with each other regularly through computer networks.

The WTG sees itself not so much an eco-cooperative, but as a group of growers with a common interest in finding solutions to environmental problems at the level of the individual enterprise. Many of the group's members, for instance, were active in the search for alternatives to methyl bromide, a soil fumigant. The accent of the Group is on the Westland area – a several square kilometre concentration of glasshouse horticulturists located just south of Den Haag. As described in a recent OECD case study (OECD, 1994), the Westland is The Netherlands' largest agglomeration of glasshouses. Glasshouse horticulture expanded rapidly in the 1960s and 1970s as new markets for fresh vegetables and cut flowers opened up throughout the Europe. The industry has come under increasing pressure since then, however, first from rising prices for natural gas, and more recently from external competition and stringent environmental requirements.

The WTG's action plan asked for:

1. permission to re-use of organic waste at the owner's enterprise through composting;

2. permission to employ non-approved pest control agents under certain circumstances (though the group undertook to reduce their overall use of, or the emissions from, pesticides, or both);

3. exemption from a rule requiring them to have a rainwater basin of 500 m^3 for every hectare under glass;

4. and more influence over research, with the aim of incorporating their specific needs into research programmes.

In its response to the WTG's proposal, the Government agreed to allow composting (Item 1), but held off on approving Items 3 and 4 pending more complete descriptions of their plans. Item 2 was rejected on the grounds that the environmental effects of using the non-approved chemical agents was not sufficiently understood.

EVALUATION

The more or less spontaneous formation of farmer-led eco-cooperatives in the early 1990s, and their subsequent evolution into laboratories of government policy, are both consistent with Dutch institutional and democratic traditions. Faced with increasing pressures from outside the sector, and an ever more complex – and to a large extent untried – set of environmental regulations, the farmers banded together into co-operative associations. From the Government's perspective, the emergence of these groups has proved a useful vehicle for mobilising farmer commitment to environmental protection, and for finding ways to shift more responsibility over the implementation of environmental policy to local communities. This latter view reflects a general shift in the Government's approach to environmental policy over the last few years.

In common with the experience of many of the other countries, the role of farm-level environmental plans is central to many of the eco-cooperatives' action plans. Three of the action plans are being used to help identify ways in which farming can be made more compatible with nature and landscape preservation. In another they are to be integrated with local physical plans. In all cases, they can be seen, in a sense, as extending the country's detailed land-use planning to yet a smaller level.

An evaluation of the environmental effectiveness of the Government's policy would be premature at this stage. Its decision to move cautiously is understandable, considering the complex legal and institutional issues involved in allowing even limited derogations from established regulations. The hope of all parties involved is that environmental performance will improve, or be achieved at lower cost, or both – a condition of all the plans is that the goals of government policy will be fulfilled. In the unlikely though possible event that the results of any of the "administrative experiments" fail to meet the original objectives set for them, the knowledge thereby gained should still help guide future government policy in this area.

6

NEW ZEALAND

BACKGROUND

Agriculture and its effects on the environment are of considerable importance to New Zealand, given the major role of agriculture in its economy. For the year ended March 1993, agriculture and related industries accounted for 14 per cent of Gross Domestic Product, and agricultural products accounted for 56 per cent of New Zealand's merchandise exports. Of New Zealand's 27 million hectares, just under 14 million hectares (51 per cent) are in agricultural use.

The relatively rapid development of New Zealand over the past century has raised a range of environmental issues, although farming in New Zealand is generally much less intensive than in other countries. Nevertheless, in some areas of New Zealand, significant soil erosion problems have resulted from removal of the natural forest cover. Serious soil degradation has occurred on approximately 2 per cent (300 000 hectares, mainly in the semi-arid inland zones) of the South Island, the result of 150 years of pastoral use. Drought and excessive livestock pressure have also been contributing factors. Feral animals such as rabbits have added to the grazing pressure, particularly earlier this century and more recently as current control policies and practices failed. Sediments, along with nutrient runoff and discharge of agricultural wastes, have contributed to water quality problems in some areas. Water allocation and protection of minimum flows is assuming increasing importance as more land has been converted to horticultural and dairy production.

In many cases, environmental problems associated with agriculture can be attributed to government policies. Until the mid-1980s, for example, agricultural support programmes encouraged over-intensive use of chemical inputs and other physical resources. Since it peaked in 1983, government assistance to New Zealand farmers has fallen dramatically, from a PSE of 36 per cent to a PSE of just 3 per cent. Price supports, tax concessions, loan subsidies, input subsidies and free advisory services have all been eliminated, and farmers now have to pay for government-provided inspection services. Remaining spending on quarantine, research and pest control is more than offset by costs imposed on agriculture through protection of input industries. In short, farmers in New Zealand receive by far the least government assistance of any OECD country.

Following the removal of subsidies the number of sheep in New Zealand declined, as did the volume of fertilisers and pesticides used on farms, and the area planted to forest increased. For some farmers with high debt levels, the combination of the government's macroeconomic reforms with the removal of subsidies caused considerable financial stress. Some left farming, others significantly reduced capital expenditure, and some depleted their financial and natural resource asset base. Since 1990, the decline in sheep numbers has continued. This has been partially offset by an increase in numbers of dairy cows, but total stock units still declined by over 8 per cent during the period 1985-1993. The more recent increase in fertiliser and pesticide use confirms that input use is sensitive to farm incomes and, by inference, the level of government assistance: if New Zealand had not removed its agricultural subsidies, input use would have been higher during the 1985-1993 period, and probably higher than it is today.

The Government's policy of regarding climatic risks has also been altered. Support will still be provided when an adverse event is beyond the ability of the local community to deal with it, but such support is provided in a manner that does not reduce individual responsibility for managing risk. Government expenditure on disaster relief has declined from an average of NZ$ 26.4 million per year during the period 1986/87-1990/91, to an estimated NZ$ 0.6 million in 1994/95.

The principle of sustainable management, which is the overall objective of New Zealand policy in this area, is embodied in the *Resource Management Act* 1991 (RMA), which replaced all or parts of 75 previous statutes and brought all activities relating to the management of land, water, air, and the coastal environment under one Act. The RMA has one overriding purpose: to promote the sustainable management of New Zealand's natural and physical resources for the benefit of present and future generations. Under the RMA, most environmental responsibilities are assigned to local government. As a result, the majority of resource management programmes relevant to agriculture are now carried out at the regional level. These include soil conservation activities, water quality monitoring and control, and pest management. Policy in this area is still evolving, however, as regional and district councils continue with the environmental planning process initiated by the RMA.

European settlers brought with them a number of plants and animals that have thrived in the New Zealand environment, proved difficult to control, and quickly developed populations to pest levels. The *Biosecurity Act* 1993 (BSA) provides for pest control operations under national or regional pest management strategies, which must specify the costs and benefits of control, who benefits, and proposed funding mechanisms. Any individual or group can propose a pest management strategy, which takes effect if it is approved by the Minister of Agriculture (for national strategies) or the relevant regional council. In 1996 the New Zealand Government has enacted the Hazardous Substances and New

Organisms Act, which will come into effect during 1997 once regulations have been promulgated. This Act will manage the harmful effects of hazardous substances and new organisms so as to enable maximum net national benefit to be achieved. Further, complementary, legislation (under development in the second half of 1996) will deal with the registration and use of agricultural chemicals.

In 1993 the New Zealand Government released a position paper on sustainable agriculture, as part of a wider policy to promote sustainable land management. The Ministry of Agriculture and Fisheries (MAF) has undertaken a facilitation programme designed to encourage the adoption of sustainable agricultural practices, and regional councils are also promoting sustainable agriculture as part of their responsibilities under the Resource Management Act. During the meantime, over 60 farmer-based community groups have formed to address issues connected with sustainable agriculture in New Zealand. Some of these groups include non-farmers. Some groups have been instigated by, and receive administrative or financial support, or both, from a regional council or other external source; others are entirely independent. Meanwhile, various industry groups have become involved in promoting sustainable agriculture through activities such as drawing up codes of good practice, and developing environmental strategies to address the entire life-cycle of the industry's products. As illustrated below, these groups vary considerably in size, interest, and governance. In contrast with Australia, the central government's policy so far has been to encourage and facilitate their formation but otherwise to become as little involved in these groups as possible, and no nation-wide programme exists to support them.

GENESIS OF FARMER GROUPS WORKING FOR SUSTAINABLE LAND MANAGEMENT IN NEW ZEALAND

Many trace the origins of farmer-led environmental groups in New Zealand to the Rabbit and Land Management Program (R&LMP), which ran from November 1989 through June 1995. Although the R&LMP did not originate as a farmer initiative, *per se*, it was an important factor in spurring the formation of farmer groups, and is seen as the progenitor of community based action to deal with problems of sustainability in the rural area.

The R&LMP was set up following a recommendation by The Rabbit and Land Management taskforce that a five-year programme be established to elucidate the complex economic, biophysical, economic, social and institutional interactions related to sustainability in rabbit-prone tussock grassland areas. It was jointly funded by landholders, regional government and central government. Over its nearly six years of existence, the Central Government invested NZ$ 22.5 million (US$15 million) in the programme. Of this amount, NZ$ 16.35 million was used to cover grants (via regional councils) to landholders and monitoring of the outcomes of the programme (approximately NZ$ 0.30 million (US$0.20 million) per annum). Funds were also allocated for administration and facilitation (approximately NZ$ 0.32 million (US$0.21 million) per annum) and for research (approximately NZ$ 0.95 million (US$0.63 million) per annum).[18]

The underlying approach of the programme was that landholders, together with regional councils, would each develop a "whole property" plan to address a variety of needs, but particularly those related to managing rabbits. Once approved by the MAF, each farm plan would form the basis on which assistance to farmers towards rabbit eradication and prevention would be provided. Objectives and accountability spelled out in the plan were registered against property titles. An annual assessment of each property plan was carried out in addition to intensive monitoring of the physical, economic and social outcomes of the programme. Involvement in the programme was entirely voluntary, but clearly there was a strong financial incentive to become involved.

From the outset, it was recognised that the programme was not going to rid the affected area of the rabbit pest entirely. Rather, its objective was to facilitate development of a community committed to changing behaviour toward a wide range of resource issues fundamental to sustainable management of dry tussock grasslands, including of rabbits. Were this achieved, it was argued, the community would be better placed to manage the issue themselves at the end of the programme.

In drawing lessons from the programme Elliott and Anderson (1995) stress the process by which the different individuals and groups attempted to achieve these policy objectives. First, the approach adopted attempted to consider the affected land as a system, focusing not only on physical but also socio-economic and institutional factors. In particular it identified the need for decision support systems and information to influence change to farming practices. The use of farmer-based information through experience and learning was utilised in the decision support system. Community-based stakeholder forums were established at the outset to help chart directions, contribute to actions, build consensus, strengthen relationships and trust, and share in learning. These forums were seen as the key to the whole programme.

In order to fill in gaps in information, targeted research was carried out by groups of scientists located within the communities. Control of their activities was increasingly divested to locals via the work of the stakeholder forum. A timeframe of five years was established in order to allow communities time to adjust. Finally, facilitation of community initiatives on an ongoing basis was initiated and supported, for example through the Rural Futures Trust. This community change model is now being assessed for application to other resource management issues in other parts of the country (see Box 2).

Box 2. **The R&LMP: a model for encouraging community action**

In reviewing New Zealand's five-year Rabbit and Land Management Program, Elliot and Anderson (1995) set out a generic model for encouraging greater community involvement in land management issues, which in their view captures what the participants in the R&LMP had learned from the experience. They are:

- Evaluate the historical sequence of events (technical, sociological and institutional) that have led to the current problems.
- Establish stakeholder forums and chart the future in terms of new partnerships.
- In order to meet immediate needs and reduce anxiety, focus on the most obvious technical issue or problem first. Only thereafter move into the underlying issues, and then within a systems context.
- Foster partnerships between farmers and scientists.
- Develop decision support systems based on information provided by both professional scientists and farmers. Extend the scope of the decision support systems from the farmers to their communities, and then from the communities to the region.
- Undertake experiments and trials to address the knowledge and information gaps identified through the decision support system. Locate willing researchers within communities wherever possible. Validate the data.
- Provide ways to extend learning experiences from individual farms to the community; help to encourage community leadership.
- In sum, do not dictate decisions; rather, empower the locals.

EXAMPLES OF FARMER-LED INITIATIVES

Project FARMER

Project FARMER stands for Farmer Analysis of Research, Management and Environmental Resources. This project reflects the way many current and future issues pertaining to agriculture and the environment will be addressed in New Zealand. It was developed by a group of innovative farming leaders who questioned the quality of land management advice then available to them. Their idea was to transfer the use of computer-based, dynamic modelling tools (decision support systems, or DSSs) from scientific laboratories to the homes of farming families.

The project was initially promoted to assist farm families in the South Island High Country establish more effective and systematic approaches to problem solving, so that they could build more robust farm businesses, taking into account each of the ecological, economic and social pressures and influences on their farming operations. The proponents of the project firmly believed that while legislated standards of ecological or environmental quality were necessary, by the time they were applied at the farm level the risk was that the land will have suffered. Moreover, they believed that if farm families themselves collect and use ecological monitoring information alongside their production and financial information, the results would be relevant to that land, that business and the farming family's community. The programme to date has focused on:

- promoting awareness of the potential uses of computer technology for farm families;
- developing data collection and analysis skills and techniques, with a particular focus on environmental monitoring;
- identifying knowledge gaps and working with researchers to bridge these gaps; and
- developing a network so that those with particular skills or interests can help others in the community.

Farmers trained in the use of DSSs provide introductory and advanced workshops for fellow farmers. These workshops introduce the farm families to the relevance and benefits of environmental monitoring; provide the specific skills and information necessary to use the range of computer packages available to address the management (including environmental) options for their property; and ensure that farmers understand the information coming out of relevant research projects so that they can better apply these results to their own situation, using the DSS as a tool for doing so. Training of the farmers who act as mentors is being done in conjunction with the research agencies and councils involved in the project. Farmers are also becoming involved in the development of ecological models for use in farmer DSSs. This component of the overall project is a multi-step process involving landcare groups, research organisations, and resource management agencies. It builds on the internationally recognised modelling techniques for monitoring and adaptive management known as ISPD, which are currently being developed by Landcare Research Ltd. for use in pastoral systems in New Zealand.

The Rural Futures Trust

This initiative followed on from that outlined above in Project FARMER. Having begun that project, the farmers involved found that the institutional structure for managing that, and other potential related projects which were being proposed, needed strengthening. In 1994 they set up The Rural Futures Trust, whose mission was to promote sustainability of rural land and communities by encouraging community groups to develop appropriate projects to "bridge" their information and knowledge gaps. The main function of the trust is to provide a vehicle through which funding can be attracted, and through which it can be distributed and targeted at worthy farmer-led projects. In particular it enables farming communities to raise funds without having to set up the necessary legal framework often required by funding bodies. It also hopes to provide opportunities for improving the skills base of rural communities, in an environment within which they are comfortable, and to develop partnerships with research and resource groups.

Landcare groups

Landcare groups are groups formed by land occupiers (often including other community stakeholders) which have a common interest in environmental issues relating to land and water resources, and in developing practical approaches to address those issues. The majority of landcare groups have developed in the South Island high country (where some landcare groups have been operating for as long as 30 years), but increasing numbers are now being established in lowland areas, and in the North Island. Currently there are more than 60 landcare groups throughout New Zealand. These generally are small, informal groups, often involving the whole farm family and other members of their community. The Federated Farmers itself has recently convened a meeting of diverse interests from farming, environmental NGOs, research establishments, local and central government, etc., to discuss ways by which landcare groups can be further fostered.

A few examples of initiatives illustrate the range and diversity of activities of these landcare groups:

- While action against major pests is undertaken under the national umbrella of the Biosecurity Act, there are many local initiatives for collective action on pests. Groups have been formed to address land and water monitoring and sustainability issues in the South Island high country, focusing on eradicating weeds and pests (*e.g.*, N*assella* tussock, broom, possum, rabbits), and on monitoring and managing other species, such as H*ieracium* weed.

- The Women's Division of the Federated Farmers has prepared a Landcare Action Guide, which has been widely distributed to "grassroots" existing and potential landcare groups, territorial authorities and other interested agencies. This Guide provides a step-by-step kit for setting up and maintaining a landcare group.

- Three "rivercare" groups have been established in Golden Bay, South Island (Figure 10). Farmers have recognised the importance of the rivers and their water and have come together to learn more about the river environment and to act collectively to maintain and enhance it. Their initial focus has been on the technical aspects of river protection work. Their

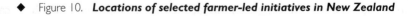

◆ Figure 10. **Locations of selected farmer-led initiatives in New Zealand**

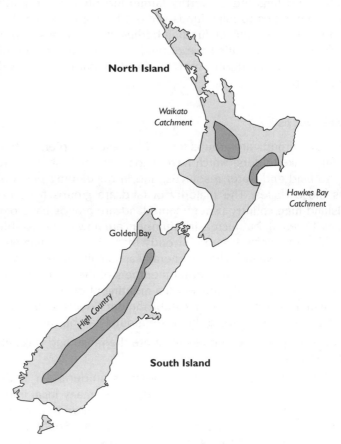

Source: OECD Secretariat.

activities have been broadened to act as a forum for discussion of land and river resource management issues, and a means to raise awareness and understanding of river resources.

- A group of 28 individual property owners along Spring Creek, Blenheim have established the Spring Creek Waterways Association to encourage landowners to preserve and enhance the natural features of Spring Creek and its margins. All enhancement activities are funded by landowners alone who are mainly farmers, but also include other members of the community. The creek originates from a spring, is around 10 km long and has high recreational values for anglers. These values are being enhanced by planting trees, reducing soil erosion and runoff from land used for intensive cropping, fencing, and increasing individual awareness of chemical runoff and the like.

- A national landcare newsletter, *Landlink*, has been initiated by farmers. It is published quarterly and aims to share information gained from landcare groups around New Zealand in order to avoid duplication of effort and to learn from others' experiences. (*Landlink* can also be accessed through the Internet at the following URL address: http://www.landcare.cri.nz/landgrps/landlink.htm)

The New Zealand Dairy Research Institute

The New Zealand Dairy Research Institute (DRI), which is affiliated to the Dairy Board, is one of New Zealand's largest research institutes. One of the DRI's more innovative projects has been to fund a three-year study of the effects that dairying is having on the quality of water in a predominantly dairying catchment, in the region of Waikato. Part of this study involves the identification of farm management practices that would lead to an improvement in environmental quality. It emerged partly as an alternative to the local Regional Council's Transitional Regional Plan, which many dairy farmers in the region regarded as inappropriate. The Waikato Federated Farmers convinced the council that farmers themselves needed to take a pro-active role in land management issues and to look for ways of promoting practices from a sustainable farming position. In 1994 they initiated a three-step process intended ultimately to produce a guide of management practices that could promote good environmental outcomes.

As a first step, the farmers, in co-operation with AgResearch Ltd and the Waikato Regional Council, convened a series of workshops to define, in farmers' terms, an operational definition of sustainable farming. Once that was accomplished, the next step involved drawing up a list of management practices which farmers have identified as leading to sustainable agriculture. A small sample includes: managing stocking rates; managing for pasture quality; optimising the use of fertilisers as part of an overall nutrient management strategy; and controlling weeds and feral animals. The farmers then identified how these improved management practices could be achieved and monitored and agreed to a set of indicators that farmers could use to describe where land, animals, and water sit on a scale of sustainability. For each of these indicators the farmers then identified the factors through which their goals would be achieved. For example, productive pasture would be achieved by having high yielding pastures, vigorous pastures, quality pasture, fertile soils, and well structured soils.

The farmers recognised that they were already using a range of monitoring tools for measuring sustainability. For example, they were already keeping records on their herds (numbers, production, health), on the conditions of their pastures, and on weather conditions and finances. To these they added a number of other indicators, particularly on the conditions of their animals and of their pastures. In the main, they chose indicators that could be verified visually, such as the condition of their animals' coats, and the weed content of their pastures. Their findings were published recently in a Resource Manual on Sustainable Farming, which contains guidelines on the management principles and methods that farmers themselves see as essential in their continuing pursuit of sustainable farming.

The next stage of the process was to design indicator scales that farmers could use themselves, with their own measurement systems, to gauge the state of their resources and progress being made. These scales were designed to enable farmers to better manage their resources and to determine whether a certain farming practice contributes or otherwise to the sustainability of the farming system. During further workshops, the farmers designed indicator scales on a (0) unsustainable to (20) sustainable level. Scales for animals, vegetation, soil and water were produced using subjective observational information (Figure 11).

While subjective scales were considered adequate for use in evaluating individual properties, they were considered inappropriate for comparing conditions among different properties. For one, farmers' beliefs about the status of natural resources will change over time and this change can only be measured if the subjective scales are quantified with objective indicators. The next part of the project aims therefore at transforming the subjective scales to ones based upon objective criteria. A workshop will now be held with 30 farmers to develop scales related to market access, family health, rural services and control of feral pests – components of sustainable agriculture developed by farmers at previous workshops. The results of use of the resource scales by farmers in study groups will be compared with scientific results to resource assessment over a three-year period.

The New Zealand Meat Research and Development Council

Many farm businesses in New Zealand stand to benefit from better planning.[19] To help overcome this problem, the NZ Meat Research and Development Council (MRDC), the producer-funded research subsidiary of the NZ Meat Producers Board, established a Monitor Farm Program in 1991. Its main aim is to demonstrate to sheep and beef farmers how farm business planning and monitoring can be used to set and achieve goals, such as measuring and managing the factors affecting bottom line profits, and managing risk. There are 23 monitor farms spread throughout New Zealand, representing a cross section of sheep and beef

◆ Figure 11. **Waikato farmer assessement indicator scales**

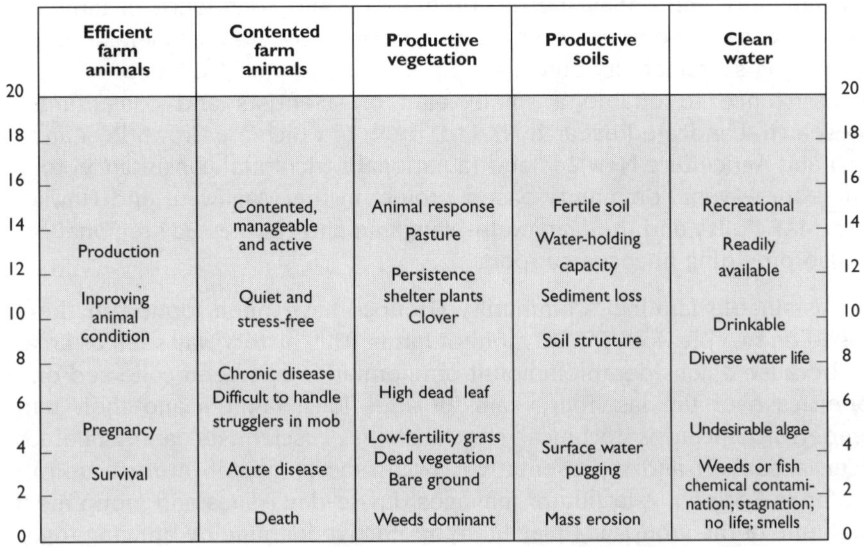

	Efficient farm animals	Contented farm animals	Productive vegetation	Productive soils	Clean water	
20						20
18						18
16						16
14	Production	Contented, manageable and active	Animal response Pasture	Fertile soil Water-holding capacity	Recreational Readily available	14
12		Quiet and stress-free	Persistence shelter plants Clover	Sediment loss		12
10	Inproving animal condition			Soil structure	Drinkable	10
8		Chronic disease Difficult to handle strugglers in mob			Diverse water life	8
6	Pregnancy		High dead leaf		Undesirable algae	6
4	Survival	Acute disease	Low-fertility grass Dead vegetation Bare ground	Surface water pugging	Weeds or fish chemical contami- nation; stagnation; no life; smells	4
2						2
0		Death	Weeds dominant	Mass erosion		0

Source: Adapted from NZ-MAF (1995).

farms. The farms to be monitored are chosen co-operatively by local agribusiness communities. Each community group is composed of up to 40 people – including farmers, veterinarians, advisers, scientists, financiers, processors and other agribusiness experts – and is co-ordinated by a facilitator contracted to the MRDC.

A key component of the Monitor Farm Program is the development of a farm business plan in consultation with the community group, who meet regularly to discuss it and review results. Through their participation in this planning process, the group members benefit by learning to analyse information and evaluate options, while the monitored farm in turn benefits from their collective experience. The wider farming community in the area is also involved through regular field days where information is available and results are presented. The programme has also established monitoring systems for key indicators, including: pasture growth and cover, soil testing, animal health status (drench resistance, disease incidence, etc.), livestock performance indicators (e.g., lambing percentages, weaning weights), and financial returns.

Overall the programme is considered by those involved to be an extremely successful vehicle for transferring practical on-farm technology to the wider farming community and enhancing the profitability and confidence of farmers. This approach was extended in 1995 on a pilot basis, to assess its general applicability to achieving sustainability outcomes. Specifically, the MRDC has provided financial assistance to enable a small team of scientists and consultants from AgResearch, Landcare Research NZ Ltd. (both of which are Crown Research Institutes) and Agriculture New Zealand (a national agricultural consultancy) to implement a three-year community-based project in the Manawatu and Hawkes Bay areas. MAF Policy and the Manawatu-Wanganui and Hawkes Bay regional councils are also providing financial support.

Sustainable farming "Community Groups" have been formed in the areas, focused on two previous MRDC monitor farms. This history was seen as an advantage because a considerable amount of information has been collected on these properties over the last four years. Besides local farmers and their partners, membership includes technical specialists (e.g., scientists, a regional council officer, a banker) and representatives from environmental interest groups (e.g., the Maruia Society). A facilitator manages day-to-day issues and group meetings. The theme of the groups is to facilitate interactive learning by bringing together a diverse range of views, while maintaining a dominant farmer flavour at all times. The desired outcomes include improved understanding by Community Group members of issues involved in sustainable pastoral farming, and development of a workable method which can be used by other Community Groups. The objective is to develop several alternative land use scenarios with a range of probable socio-economic and ecological outcomes. The Community Groups are to meet at intervals determined by the members, 4 to 6 times a year. Technical work – including the construction of physical resource inventories and the drawing up of whole-farm management strategies – continues between meetings. The process will be evolutionary, with each stage subject to Group input and ratification.

EVALUATIONS

A survey was undertaken in early 1995 which sought input from regional and district council staff and farmer representatives throughout the country on the factors that they considered to have most assisted in the acceptance of sustainable land management practices (New, 1995). Respondents to the survey showed a strong preference for the autonomous or *bottom up* approach to sustainable land management, one driven by farmer-led groups rather than imposed by some external agency. They also recognised that some groups needed assistance at the beginning to ensure motivation during the crucial initial stages of their operation. Many respondents noted that groups develop at different speeds, with a small number of groups providing most of the initiatives. The result is a highly diverse range of sustainable land management initiatives, of particular relevance to the local conditions.

Local authorities and branches of the Federated Farmers with active landcare groups perceived these groups as being an excellent way to promote sustainable land management, providing a forum for developing co-operative and often innovative solutions to land management issues. Regional councils viewed their own role as one of facilitating, encouraging and liaising with the landcare groups, as well as providing the groups with assistance and information where the local authorities are unable to do this. Regional councils pointed to the need to present information in a form that is accessible to all stakeholders. Their suggestions for improving the ability of farmers to efficiently and fully search scientific literature included upgrading library resources and electronic networks, and encouraging communication between research organisations and land users through research and technology transfer partnerships.

Respondents to the survey saw willingness to alter traditional land use methods and practices as an important issue in implementing sustainable land management. Perceptions about the degree to which land users have in fact demonstrated their willingness to adapt differ, however. Respondents to the telephone survey, such as the regional councils and the Federated Farmers, portrayed land users as becoming increasingly proactive – i.e., beginning to identify potential problems before they occur. (There was no reiteration of this view in replies to the written questionnaire.) All regional councils responding to the telephone survey identified ongoing consultation with land users as essential for the successful promotion, and subsequent attainment of, sustainable land management practices. Follow-up forums and meetings were seen as important components of this ongoing consultation.

Constraints to addressing environmental issues in agriculture were also identified. Local authorities (as well as farmers) complained that insufficient capital was available to initiate programmes, or to further develop existing programmes. The reluctance of land users to accept any form of intervention from any level of government was also cited. Some land users, they felt, refused to accept that sustainable land management is needed, or questioned the validity of separating sustainable land management from traditional soil conservation practices and methods of implementation. Finally, they felt that groups and organisations sometimes pushed personal agendas rather than striving for a common cause as a collective.

The farmers, for their part, regretted the competing priorities that result in them having insufficient time to dedicate to sustainable land management initiatives. They also complained of a general lack of information concerning sustainable land management as well as a lack of non-contestable scientific studies indicating how land use practices impact negatively on the environment. A lack of consensus amongst stakeholders concerning the meaning and practical implications of sustainable land management was also inhibiting progress. Some regional councils and branches of the Federated Farmers said they needed more staff. The general perception was that there was a role for someone to actively promote and encourage sustainable land management as well as to provide advice, information and education to land users. The Federated Farmers also claimed that there are still too few farm demonstrations or model farms to show alternatives to current practices and identify sustainable versus unsustainable land management practices.

District councils and Federated Farmers perceived problems of poor communication and poor relationships between council staff and land users. Eight local authorities believed that there was a lack of commitment from land users and the community to accept that sustainable land management was needed. Regional councils noted the poor financial situation of land users, aggravated by low prices for their products; the lack of monitoring and measuring effects of land management practices and identification of useful and practical indicators; and the failure to identify what the role of the regional council should be in sustainable land management.

The territorial authorities observed that their organisational structure can inhibit co-ordination and communication between different departments, and that this in turn can inhibit a co-ordinated approach to developing and implementing sustainable land management programmes. They complained also of the lack of political will from councils to seriously tackle the issue of sustainable land management. Finally, they noted that questions of equity – *e.g.*, between beneficiaries of sustainable land management and land users, including costs of implementation – remain contentious.

RECENT DEVELOPMENTS

In recent years the infrastructure and institutional arrangements required to achieve sustainable land management across the country have been improved through the combination of legislative, policy and economic reform. The combination of these reforms and the co-incident community based approach to dealing with current land management problems have resulted in substantial changes in attitude and in land management practice. The initiatives outlined above in this short paper are but a small sample of the large number of initiatives currently underway throughout the country.

In response to the results being achieved through this community based approach the New Zealand Government has recently refocused its approach to stimulating further progress. As part of a broader environmental policy the

Government has given priority to the development of a Sustainable Land Management Strategy for New Zealand (Ministry for the Environment, 1996). The purpose of the Strategy is to enable land users, and those who provide support and services to them, to work together more effectively and to add value through the sharing of resources, information and results. It is intended that the Strategy will provide the framework from which initiatives on the part of industry sectors, local government, communities and land managers, along with the programmes of Government agencies, may be better aligned. In this way, long-term and permanent changes in management practice will be achieved through applying a model of continual improvement. The provision of advice and support to land managers and community based groups, and the co-ordination of support systems such as research and technology transfer, will have the effect both of adding value to current initiatives such as those described above and in stimulating new initiatives to deal with current problems.

In parallel to the Sustainanble Land Management Strategy, in early 1996 the Government established a new National Science Strategy (NSS) for Sustainable Land Management (Ministry of Research, Science and Technology, 1996). In taking this initiative the Government explicitly recognised that land users – both public and private – must play a central role in setting research priorities directed towards providing better understanding of current land management systems and in providing information from which improvements can be achieved. The NSS will be implemented by a central committee and three regional NSS Committees. The central committee will draw its membership from the science community, commercial land users and environmental and other non-governmental organisations, as well as government agencies. The regional committees (one on the South Island and two on the North Island) will have no restrictions placed on membership numbers, though they will be expected to achieve a balance of scientists and sectoral interests. The Strategy is expected to strengthen the links between science, policy and land management, particularly in the transfer of the findings of research to land managers and community-based groups.

7

SUMMARY AND CONCLUSIONS

"e pluribus unum"
Out of many, one

That there exists some potential for community-based partnerships to stimulate innovative, low-cost yet effective approaches to sustainable agriculture would appear to be borne out by the growth of such voluntary initiatives in several OECD Member countries. The present study has examined the experiences of groups in Australia, Canada, The Netherlands, and New Zealand, but the phenomenon trend is not confined solely to these countries.

The aim of this final chapter is to search for common lessons from these disparate examples and, from these, to derive some preliminary conclusions of use to policymakers. The first section examines the salient characteristics of the groups and their main concerns and activities. The second examines the role of government policies to promote them, and the general agricultural and environmental policy context in which they apply. The third and final section concludes with some suggestions that could be of value to policymakers.

CHARACTERISTICS OF FARM COMMUNITY GROUPS

No social group forms in a vacuum, and farmers' groups are no exception. As discussed in Chapter 2, analysis of the motivations, structure and activities of different groups can help in the identification of the combinations of characteristics that are favourable to collective action. The literature (*e.g.*, Olson, 1971) suggests any number of characteristics that can have a bearing on the cohesion of groups and the effectiveness of their leaders. Examined below are those for which it was possible to obtain information. The degree of heterogeneity within groups with respect to size of farm or income, a factor that might have a bearing on group behaviour, was not examined in this study.[20]

Size and composition

Typically, the farm community groups examined in this study range in size from 20 to 100 members, with many around the mid-point of this range. There are a few exceptions, with some groups having memberships as large as 1 000, but even in these most activities are carried out by smaller sub-groups. Group membership is typically drawn from local communities or is based on natural units, such as small watersheds. Some groups are composed exclusively of farmers, but the trend seems to be to open membership to other interested individuals from the same communities. Generally, the groups are composed mainly of people engaged in the same type of farming activity.

Motives behind the formation of farmers' groups

All else equal, one would expect that motives that involve long-term values and needs would favour collective action over other, more short-term motives. The emergence of voluntary farm community groups in the several OECD countries examined appear to have been prompted by one or more of the following motives:

– concern about declining farm profitability;

– increasing awareness of the linkages between certain aspects of ecological sustainability and farm financial sustainability;

– fear that solutions to problems of sustainability in general, and pollution in particular, would be imposed by a central authority, combined with confidence within the groups' membership that by taking their own initiative they would be more likely to achieve satisfactory, locally acceptable results than if they were simply to wait for the government to impose a solution. (Australian farmers speak of taking "ownership" of the issues.)

In general, these three motives have played an influential role in most of the countries and regions examined. In some cases – the wheat belt of Western Australia is a good example – the first two were closely linked. Of the three, the second is most likely to provide an enduring motivation for farmers to develop more sustainable farming systems, though not necessarily always through collective action. It is, however, one that leads to a focus on issues of local importance, less so on issues that are of concern to interested parties outside the community in which the group operates. The third is perhaps the easiest for governments to influence: an implicit understanding that if local groups fail, government will step in. This motive can provide an effective, ongoing spur to collective action. But its effectiveness is critically dependent on the credibility of the understanding and the ability of government to monitor performance.

The external benefits that farmers expect to derive from co-operative action appear to fall into two main overlapping categories:

• those relating to improvements in the environment itself, such as reductions in salinity, wind erosion, pests, and improvements in habitat for wildlife (especially game and insect-eating birds);

- those relating to co-ordination and the exchange of information, which often are related to scale economies.

Both of these categories are related to protecting the value of farm assets and avoiding regulations that the farmers would regard as burdensome. Such results are consistent with recent empirical work conducted in the United States by Weaver (1996), who found that in general farmers are mainly motivated to provide public environmental goods by factors that influence profits, and only weakly by non-hedonistic values – e.g., altruism, or the "warm glow" from giving. The intangible, social benefits of group membership should not be completely discounted, however. In the remote farming communities of the Australian outback, increased social contact is often cited as a side-benefit of participation in landcare groups. More generically, the groups seem to have provided an important outlet for channelling the energies of community leaders, and to have created in their membership a heightened sense of control over their own destinies. As is clear from the case histories of the groups in Australia and The Netherlands, they have also served as local forums for discussion, thus helping to improve the relations between farmers and environmental interest groups.

Concerns addressed by the groups

The voluntary farm community groups examined have developed organically from within the community of interests involved with them. Not surprisingly, they have therefore tended to focus on local issues, especially those for which there are important externalities. At least initially, the farm community groups examined tended to put less of a priority on issues relating to water quality than soil quality. Generally, those that have concentrated on nutrient pollution of groundwater have done so either because existing regulations compelled them to (as in The Netherlands), or because of governments steered them in that direction (as in Canada). As groups that have been around for several years have matured, however, they have often expanded their scope, and begun to look at sustainable land management from an integrated, systems perspective. This change in perspective appears to have been encouraged in some cases by government efforts to link individual farm plans with plans for larger geographic units, particularly watersheds (e.g., the Murray-Darling Basin in Australia).

How effective the farmers' groups have been in tackling environmental issues is difficult to evaluate at this stage. Even in Australia, where surveys have begun to distinguish between farms whose operators are members of landcare groups and those that are not, most of the data collected so far relate to farming practices, not results. At least in this respect, the results appear to be encouraging, in as much as landcare group members tend to have completed farm management plans and adopted soil conservation practices sooner than those that are not members of landcare groups. Such results must be interpreted with caution, however, because of the problem of self-selection: those individuals that tend to be more aware of sustainability issues and to be willing to adopt new practices are also probably the same population from which the landcare groups draw their members. In other words, participation in landcare groups may to some extent reveal a predisposition to responsible land management, and not be the sole cause of such behaviour.

Elsewhere, data that would allow an evaluation of the environmental effectiveness of farmer groups is even more sparse. As in Australia, the available information on New Zealand's groups relates mainly to activities, attitudes and, to a lesser extent, farm practices. Published evaluative studies in Canada or The Netherlands are only just starting to emerge.

The ability to conduct such evaluative studies should improve in time, however. More and more farmers are completing and maintaining farm plans, and beginning to monitor their performance against indicators of economic, physical and biological conditions. These data could presumably be used in comparative evaluations. Ironically, the fear of proprietary data being used in legal action against them is sometimes seen as a reason behind some farmers' reluctance to prepare farm plans. Such fears could be allayed, however, if farmers were given assurances that the data they collect would be treated anonymously, as is done under the Atlantic Farmers' Council EFP Initiative in Canada.

As would be expected, the farmers' groups still give little attention to environmental issues important to populations outside their areas of influence. Few if any groups are addressing greenhouse gases, such as methane, though the steps some are taking to improve the management of livestock waste may be contributing to reduced emissions. And although many groups in Australia and New Zealand are working together to control *pests* (mainly vertebrate pests and weeds), few are concerned with reducing risks from the use of *pesticides*. Again, where community groups have taken up such issues, they have usually done so in response to government policy (as in The Netherlands). Clearly, there is a range of environmental issues that governments cannot expect such groups to address through self-regulation – *i.e.*, spontaneously, out of their own self-interest.

Tasks and activities

There is a great diversity in the specific initiatives taken by the farm community groups studied here: some have formed mainly to provide a forum to exchange information; others have taken more active initiatives, financed at least in part by their members. Apart from exchanging information, the most common activity, especially in the groups' formative stages, has been to work together in preparing farm plans. Generally these farm plans have taken a "whole farm approach" and have been used as aids to encourage farmers to consider all the environmental, economic and sociological factors that bear on the sustainability of their enterprises. Often, outside advisors (from government extension offices, universities or producer organisations) have been enlisted to help design the plans or to conduct workshops to help farmers learn how to use the plans. In New Zealand, model farms have been set up in a few localities to help farmers preparing farm plans through "learning by doing." Farm-level demonstrations have also been used by a number of groups to test out new, low-cost approaches to sustainable land management. The active involvement of the community groups working with the farm owners is seen as being critical to their success. They can observe at first hand the results of different management strategies and new technologies.

Many groups organise work days, during which members work collectively on particular projects, such as constructing nature inventories, repairing and maintaining earthworks, or eradicating noxious weeds. In general, the groups examined appear more inclined to undertake activities that pool labour than pool capital (except when the capital is provided by government). Difficulties in devising mechanisms for apportioning financial contributions on a voluntary basis may be a barrier. The fact that some groups have managed to develop innovative approaches to raising money, however, suggests that collective funding of activities could grow in importance as groups move from the planning and evaluation stages to implementation. The recently announced A$ 380 million (US$300 million) Murray River environmental land and water management plan in Australia demonstrates the possibilities that such approaches present.

THE ROLE OF GOVERNMENT POLICY

A fundamental characteristic of most of the groups examined in this study is their close association with their local communities. This has led to better co-operation between the groups and the administrations charged with the more general well-being of the villages, towns and municipalities in their areas. Clearly, though, national governments have influenced and will continue to influence the formation of farmer-led environmental groups and their activities – both directly, through policies targeted directly at the groups, and indirectly, by influencing the economic and policy environment within which they operate. It is this policy context that is discussed below.

Policies to promote and support farm community environmental groups

Many of the differences in the level and nature of support provided by central governments to farmer-led groups reflect perhaps as much the relative newness of the groups in their countries as differences in policy approaches. In Australia, both the federal and State governments are quite actively involved in providing seed money and strategic support to their farm community groups, and policies towards them are well-developed. The Netherlands, which is at a much earlier stage, did not establish a formal policy towards its eco-cooperatives until February 1996.

Besides scale, however, one can observe differences in other respects. While all four of the countries examined provide some support to farmer-led environmental groups, either through direct aid or advice – usually both – the form of support and the level per farmer (or per group) varies considerably. New Zealand sits at one end of the spectrum: its central government allocates no funds specifically to support landcare-type activities, though it has allocated money to projects in the past (most notably the Rabbit and Land Management Program); most funds nowadays come from local governments, industry, charitable organisations and the farmers themselves. The other three countries provide funding in amounts ranging from around US$0.25 million a year in The Netherlands to US$50 million a year in Australia (including the Commonwealth-State component). The amounts involved in The Netherlands work out to approximately US$50 000 per group per year, but this is not necessarily an indication of the government's long-term policy towards eco-cooperatives, which is still being worked out. In Australia, the policy so far has been to spread whatever funds are available as widely as possible; annual direct payments per recipient group average less than US$15 000, and only about one-third of the groups receive money in any given year. The Government's view is that such payments should serve more as a catalyst to private spending than be seen as compensation; initial data appear to confirm that they are indeed having the desired effect.

A crucial element of the programme in Australia (and also in The Netherlands) is that no funds are provided directly to individual farmers, only to community landcare groups. Among other advantages, dealing with groups of 40 or 50 individuals probably reduces administrative costs and improves targeting, compared with what would be required to administer a programme that made payments to the same number of land holders directly. This conditionality on payments, though not a factor in the genesis of the landcare groups, is probably also contributing to their current growth in numbers, providing an added incentive to their formation. Whether such payments can be sustained at a moderate level as the overall participation rate increases beyond 50 per cent (assuming it eventually does), remains an important question. Again, the phenomenon of self-selection may have to be considered: those who remain outside the groups presumably would require a greater monetary incentive to join than those who are already members.

Another policy aspect that varies among countries is the degree of structure of their approaches, which is generally correlated with legal compulsion.[21] In Australia, Canada and New Zealand, commitments made by farm community groups to their responsible environmental and agricultural authorities have tended to be informal. Moreover, these commitments have in almost all cases been in respect of goals and not means, and have been expressed in general terms, such as endeavouring to attain a state of sustainable agriculture by some date in the future (without defining such a goal too precisely). Sanctions for non-attainment are generally not part of such commitments. By contrast, the situation with regard to the five "eco-cooperatives" highlighted in the chapter on The Netherlands is more formal. While not as structured as the official covenants that the Dutch Government has secured with other industries, the agreements are likely to include evaluation criteria, review schedules and perhaps even sanctions for non-attainment of goals. It is easy to see why, in a country as densely populated and as intensively farmed as The Netherlands, such legal safeguards are considered necessary.

The general agricultural and environmental policy context

Other policies – particularly agricultural support policies and environmental policies – have undoubtedly influenced the formation of farm groups, and their effectiveness in addressing particular problems. As mentioned above, an understanding of the linkages between ecological and financial sustainability at the farm level appears to have been a prime motivation driving many such groups. This awareness, it is often argued, is inhibited by the conflicting signals sent through agricultural support policies.

All else equal, a farmer operating in a policy environment that assures adequate returns from farming through government interventions in the market, or that provides generous disaster assistance, will probably be less concerned about the long-term consequences of practices that risk degrading his land than in the absence of such policies. Similarly, policies that discourage crop rotation or that favour the production of one commodity over another, are also likely to obscure the link between profitability and sustainable farming practices. Validating such relationships is extremely difficult, however. At most one can observe the countries in which co-operative farmer groups have developed the furthest – Australia and New Zealand – are also the countries that have consistently provided the lowest levels of agricultural support, and whose farm incomes are most dependent on movements in world market prices. Analysis of the policy situations in other OECD countries where similar approaches are emerging would shed more light on this crucial question.

Of course not all agricultural policies act as a deterrent to self-help; some would appear to facilitate it. For example, the prior existence of a well-developed "agricultural knowledge system" – a feature of all four of the countries examined – may have helped create conditions that were favourable to farmer interest in co-operative approaches. The OECD's First Joint Conference of Directors and Representatives of Agricultural Research, Agricultural Advisory Services and Higher Education in Agriculture (4-8 September 1995), lent credence to such a view, observing that, often, "the success of programmes related to sustainable agriculture and improved environmental practices were due to a close co-operation and concerted action between the three main functions of the [agricultural knowledge system], leading to new partnerships and networks." A very concrete example of this can be found in The Netherlands, where one of the farmer environmental groups (Werkgroep Telen in de Grond) has formed out of the horticultural industry's long-standing Foundation of Horticultural Study Groups.

OVERVIEW AND SUGGESTIONS OF RELEVANCE TO POLICY MAKERS

The experiences to date of community-based farmer environmental groups provides encouragement for the view that co-operative approaches to responsible management of agricultural land and water resources can play a useful role in public policy. They are by no means the only group-based structures suitable for implementing environmental policy objectives, of course. Labelling schemes, such as the *appellation contrôlée* systems that exist in several European countries, and which are based not so much on farm communities as on groups organised along product lines, have also had some success in regulating the environmental performance of affiliated producers. Both should be considered as possible approaches to be included as part of a larger package of policy measures. However, it should also be recognised that the suitability of group-based approaches may be limited by the motivation of farmers (and other interested individuals) to participate in such activities, and the environmental issues they are both willing to address and capable of addressing effectively. As shown in the theoretical discussion in Chapter II, for example, the geographical distribution of the environmental effects of farmer behaviour in relation to abatement costs is likely to have an important bearing on the potential gains from joint action.

Assuming that such groups can play a useful role in addressing certain issues related to sustainable agriculture, what might governments do to help foster their formation? Box 2 in the New Zealand chapter provides a list of practical suggestions for encouraging greater community involvement in land management issues, all of which are supported by the experiences of the other countries examined in this report. In addition to those, several more general observations that flow from the foregoing discussion might be considered:

- *Reform agricultural policies so as to eliminate conflicting signals.* Farmers producing in an environment in which a large degree of production risk has been eliminated by implicit guarantees by the government may look less closely at the consequences of their own actions on financial and ecological sustainability. Policies that distort production decisions and discourage the adoption of more sustainable land management practices also merit close attention as candidates for reform. Obviously, further analysis is needed to confirm this point.

- *Make some project funding available only or principally to groups, not individuals.* Providing limited financial assistance to projects proposed by groups rather than by individuals is likely to favour activities that benefit communities as a whole rather than certain individuals. The availability of such money, in turn, is also likely to encourage greater community involvement.

- *Emphasise training and encourage partnerships between farm communities and scientists.* Farmers in many countries are naturally sceptical of collective action. Training programmes that bring farmers together in an atmosphere conducive to the sharing of information and experiences can help break down some of these barriers. Such training should be aimed at providing farmers with skills of institution building and the management of organisations, and at increasing competence and commitment to group action (Sinha, 1996). Fostering partnerships between farm community groups and scientists helps create a climate of innovation and leads to fruitful exchanges of knowledge.

- *Work with stakeholders to develop indicators and other decision support systems that are useful both to land-holders and to regional planners.* Farmers, like any other decision makers, require good data in order to make informed decisions. Involving them in the development of such systems can both improve their understanding of and appreciation for issues related to sustainability and empower them to take action. Their location-specific knowledge can also be invaluable in the design of indicators. Involving them in planning processes at the regional level – especially in rainfall catchment and river-basin planning – can help improve coherence between farm-level actions and regional initiatives.

- *Create an enabling environment for the devolution of responsibility.* The formation of self-organised, farm community environmental groups should not be seen as a policy goal in itself. Rather it should be viewed as part of the broad, society-wide quest for subsidiarity: the devolution of responsibility to the smallest competent political unit. This does not mean that such groups can be relied upon to manage all aspects of all issues related to sustainable agriculture – many complex or transboundary issues are likely to be beyond their competence. There is therefore always bound to be a need for sub-national and national governments in monitoring performance, in establishing and enforcing standards, and in providing advice and other strategic support. But as farm communities develop their awareness of sustainability issues, and demonstrate their capacity for dealing with them, the last thing governments should do is stand in their way. This means, in many (but not all) cases, avoiding environmental policies that are overly prescriptive in respect of production methods and technologies, so as to allow producers some flexibility to adopt not only traditional but also innovative approaches for dealing with the challenges of sustainable agriculture.

"The roots of social order are in our heads, where we possess the instinctive capacities for creating ... a better [society] than we have at present", concludes Matt Ridley (1996) in his recent study of *The Origins of Virtue*. The challenge for policy makers interested in using co-operative institutional arrangements to promote more sustainable agriculture is to design them in such a way that they tap into and draw out those instincts.

Annex

JOINT IMPLEMENTATION IN AGRICULTURE: NEW YORK'S WATERSHED AGRICULTURAL PROGRAMME[22]

New York City's water supply system is one of the largest in the world, yielding 4.5 billion litres a day. In 1986 the US Congress passed a law that, unless other measures were taken, would have required New York to filter its water at an estimated up-front cost of US$5 billion to US$8 billion and annual operating expenses of US$200 million to US$500 million. Initially the city sought to avoid these costs by regulating activities in its three watersheds and acquiring additional land. Dairy farming in the Catskill and Delaware watersheds (which provide 90 per cent of New York City's supplies) would have been severely curtailed by the proposed regulations, which would have prohibited storm-water runoff from grazing areas and drainage from barnyards and feedlots from entering into any watercourse (New York City Department of Environmental Protection, 1990, pp. 27-28).

After vigorous debate among the opposing interests, the New York City Department of Environmental Protection (NYC-DEP) and the agricultural community agreed to establish an *Ad Hoc* Task Force to recommend an alternative plan of action. The Task Force concluded that it would be far better for the city to "withdraw the proposed regulations on agriculture" and to implement "a voluntary, locally developed and administered programme of best management practices" (*Ad Hoc* Task Force, 1991, pp. 4-5). A new not-for-profit corporation, the Watershed Agricultural Council (WAC), was created for this purpose, consisting of 19 farmer and agribusiness leaders from across the watershed region, the Commissioner of the NYC-DEP, and 11 non-voting members drawn from government and private organisations.

The centrepiece of this new experiment in watershed management is a comprehensive "Whole Farm Planning Program", which seeks both to protect water quality and to strengthen the economic viability of farming in the affected watersheds. In order to meet the strict quality criteria for New York City's unfiltered drinking water system, the plans include strategies to deal with a wide range of potential pollutants, including animal pathogens such as the protozoans *giardia* and *cryptosporidium*.

The programme is being implemented in two phases. Phase I, which began in 1992 and ran for two years, concentrated on developing and implementing whole farm plans at ten demonstration farms, and on training Country Project Teams (comprising experts from local Soil and Water Conservation Districts, Cornell University's Co-operative Extension office, and the USDA's Natural Resources Conservation Service) in each of the eight counties within the watershed. County Project Teams, co-operating farmers and a technical support group devised the plans; the WAC then reviewed the plans and approved all of them wholly or partially. The costs of implementing them was fully covered by New York City, up to a total cost of US$1 million. In all, around US$5 million was spent in Phase I.

The basic aim of this first phase of the project was to systematically gather data and develop scientific foundations and inter-organisational ties so as to increase the likelihood of success in the full-scale Phase II, which ran from 1 October 1994 through the end of 1997. Among the primary objectives in the programme's second phase has been "to foster community pride, enthusiasm, and empowerment through local leadership and involvement in a nationally recognised, innovative, co-operative approach to a highly complex environmental situation". A target has been set to encourage at least 85 per cent of the watershed's 500 farmers to develop whole farm plans and to implement best management practices for preventing pollution. If that target is not met, then the NYC authorities and the WAC will decide what if any changes in the programme may be needed in order to protect the city's water supply from agricultural pollution. In such an event, the possibility of introducing new regulations could be considered. Phase II is funded at US$35 million.

NOTES

1. Following Glachant (1996, p. 6) an institutional arrangement "characterises the way that [policy] transactions are organised". That is to say, "it describes the interactions of the role of the agents involved in policy making (*i.e.,* the regulator and the regulatees)".

2. The term "voluntary" is sometimes also associated with approaches that attempt the reverse – that is, which subsidise changes in behaviour in the hope of influencing attitudes – in that the recipients of such assistance are not forced to accept the payments they receive. This is not the sense of the word meant here.

3. The Farm Films Recovery Scheme, which was launched in 1995 and managed to collect and recycle more than 8 000 tonnes of used silage films over the course of two years, was not a total success. Because a few farm plastic importers refused to participate in the scheme, participating companies were left with significant amounts of unsold, levied products which could not compete with unlevied products produced by the non-participating companies. The participating companies, organised under the Farm Film Producers' Group, considered this to be an unacceptable commercial pressure and in March 1996 discontinued the levy. All collections of film under the scheme were suspended at the end of February 1997.

4. Areas of low-lying land reclaimed from the sea and surrounded by dikes.

5. See, for example, OECD (1997c).

6. See, for example, note 8 in Stranlund (1995).

7. That is, assuming that the underlying goals of the measure are regarded as legitimate. A strict ban on an activity can be quite conclusive in the short run – assuming it can be enforced. But as attempts in some countries to ban alcohol consumption have demonstrated, the effectiveness of such measures can diminish over time if they are not supported by changes in personal preferences. Most countries now try to limit the abuse of alcohol by disseminating information on the hazards of over-consumption, and by taxing its sale.

8. Salient characteristics of damage functions include: whether there exists a threshold value below which environmental damage (benefit) associated with the variable of interest is negligible, or above which it is critical; whether temporal variability is high (are policy makers concerned with peak values?); and whether or not the environmental damage (benefit) is reversible.

9. See, for example OECD (1997a).

10. The area encompassed by the SWQI is defined by its predominant vegetation, a small tree known as "Mulga" (*Acacia aneura*).

11. At an average cost of A$ 64 000 (US$48 000) per property; see Anthony Hoy, "Vanishing artesian water stock triggers concern in the outback", *The Sydney Morning Herald*, 15 May 1996, p. 4.

12. Based on "The South West Queensland Initiative – A Case Study in Regional Land Management", 8 September 1995, World Wide Web (http://www.erin.gov.au/portfolio/esd/csd95/case5.html).

13. Based on "The Murray-Darling Basin Initiative – A Case Study of an Integrated Approach to the Planning and Management of Natural Resources", 8 September 1995, World Wide Web (http://www.erin.gov.au/portfolio/esd/csd95/case6.html).

14. Some of the changes, such as in salinity, are expected to be mitigated or reversed over several decades, hence unambiguous indications of change may be difficult to obtain.

15. The OFEC has benefited from support from several Ontario Ministries, Agriculture Canada, Environment Canada, the University of Guelph, and several voluntary organisations in this work.

16. These groups were: Vereniging Eastermar's Lânsdouwe; Vereniging voor Agrarish Natuur- en Landschapsbeheer Achtkarspelen; Milieucöperatie de Peel; tuinderswerkgroep Telen in de Grond; and Stichting Waardevol Cultuur Landschap Winterswijk.

17. The main sources of information for this section are the studies by Hees, Renting and de Rooij (1994), and by van Broekhuizen, *et al.* (1997).

18. A goods and services tax (GST) is levied on these expenditures, payable to the Treasury.

19. According to the Meat Research and Development Council.

20. Research in this area might shed light on the relative burdens borne by different group members. See, for example, Olson and Zeckhauser (1970).

21. Solsbery and Wiederkehr (1995) provide a typology of voluntary approaches along these lines.

22. This description is based on a paper by Coombe (1996).

BIBLIOGRAPHY

Ad Hoc Task Force (1991), "Policy group recommendations", Ad Hoc Task Force on Agriculture and the New York City Watershed Regulations, New York, United States of America.

ALEXANDER, Helen (1995), A Framework for Change: The State of the Community Landcare Movement in Australia, The National Landcare Facilitator Project Annual Report, Department of Primary Industries and Energy, Canberra, ACT, Australia.

ANDERS NORTON, N., T.T. PHIPPS and J.J. FLETCHER (1994), "Role of voluntary programs in agricultural nonpoint pollution policy", Contemporary Economic Policy, Vol. 12, No. 1, pp. 113-121.

ARROW, Kenneth J. (1977), "The organisation of economic activity: issues pertinent to the choice of market versus nonmarket allocation", in Public Expenditure and Policy Analysis (R.H. Haveman and J. Margolis, eds.), 2nd edition, Rand McNally, Chicago, Illinois, USA.

Australian Nature Conservation Agency; Commonwealth Department of Environment, Sport and Territories; and Department of Primary Industries and Energy (1995), "Landcare information: land, water and vegetation programs – 1995-96", Third Edition, World Wide Web [http:/www.erin.gov.au/land/landcare/landcare–new2.html], version of 8 June 1995.

AXELROD, Robert (1984), The Evolution of Co-operation, Basic Books, New York, United States of America.

BAUMOL, William J. and Wallace E. OATES (1988), The Theory of Environmental Policy, 2nd ed., Cambridge University Press, Cambridge, United Kingdom.

BLUM, A. (1991), "What can be learned from a comparison of two agricultural knowledge systems? The case of the Netherlands and Israel", Agriculture, Ecosystems and Environment, Vol. 33, No. 4, pp. 325-339.

BNA [Bureau of National Affairs, Inc.] (1995), "Voluntary accords seen as way to protect environment while remaining competitive", International Environment Reporter, 26 July 1995, pp. 585-588.

BOONEKAMP, W. Loek (1992), "L'organizzazione del l'agricoltura in Olandia", in Rapporto 1992 Su l'Agricoltura Italiana (R. Prodi, G. Armadei and P. De Castro, eds.), pp. 213-263, Societa editrice il Mulino, Bologna, Italy.

CAMPBELL, Andrew, with Greg SIEPEN (1994), Landcare, Allen and Unwin, St. Leonards, NSW, Australia.

CHAPMAN, Lisa (1997), "Landcare – links between Landcare, training and farm management practices", in *Australian Farm Surveys Report 1997*, pp. 34-37, Australian Bureau of Agricultural and Resource Economics, Canberra, Australia.

COMMONS, John R. (1950 [1970]), *The Economics of Collective Action*, 2nd ed., University of Wisconsin Press, Madison, Wisconsin, United States of America.

Commonwealth of Australia (1991), *Decade of Landcare Plan: Commonwealth Component*, AGPS Press, Canberra, ACT, Australia.

Commonwealth of Australia (1995), *Australia's Report to the UNCSD on the Implementation of Agenda 21*, AGPS Press, Canberra, ACT, Australia.

COOMBE, Richard I. (1996), "Watershed protection: a better way", in *Environmental Enhancement Through Agriculture*, Proceedings of a Conference, Boston, Massachusetts, 15-17 November 1995 (W. Lockeretz, ed.), pp. 25-34, Tufts University, Medford, Massachusetts, United States of America.

CURTIS, Allan and Terry De LACY (1996), "Landcare in Australia: does it make a difference?", *Journal of Environmental Management*, Vol. 46, No. 2, pp. 119-137.

van DIJK, G. (1990), *Is de tijd rijp voor milieucoöperaties?*, Rijswijk, The Netherlands.

DOKTER, Henk (1995), "Milieucoöperatie is er klaar voor", *Boerderij*, Vol. 81, No. 4, 24 October 1995, pp. 14-17.

DONALDSON, Sheila (1995), "Farm-level perspective of constraints to dealing with Landcare", in *Outlook '95: Proceedings of the National Agricultural and Resources Outlook Conference, Canberra, 7-9 February 1995* (Vol. 1: Commodity Markets and Natural Resources, pp. 190-197), Australian Bureau of Agricultural and Resource Economics, Canberra, ACT, Australia.

DOUGLAS, Jock, Helen ALEXANDER and Brian ROBERTS (1995), "Sustaining the agricultural resource base: community Landcare perspective", in *Sustaining the Agricultural Resource Base* (K. Goss et al., eds.), pp. 68-76, AGPS Press, Canberra ACT, Australia.

DPIE [Department of Primary Industries and Energy] (1995a), *Australian Rural Policies: Considerations for the 1995 US Farm Bill*, Canberra, ACT, Australia.

DPIE (1995b), *National Landcare Program: Report on Operations of the Land and Water Elements – 1993-94 Financial Year*, Canberra, ACT, Australia.

DPIE (1995c), *Managing for the Future – Report of the Land Management Task Force*, Canberra, ACT, Australia.

DUMANSKI, J., L.J. GREGORICH, V. KIRKWOOD, M.A. CANN, J.L.B. CULLEY and D.R. COOTE (1994), *The Status of Land Management Practices on Agricultural Land in Canada*, Technical Bulletin 1994-3E, Centre for Land and Biological Resources Research, Ottawa, Ontario, Canada.

EILERS, R.G., W.D. EILERS, W.W. PETTAPIECE and G. LELYK (1995), "Salinization of Soil", in *The Health of our Soils – Toward Sustainable Agriculture in Canada* (D.F. Acton and L.J. Gregorich, eds.), pp. 77-86, Agriculture and Agri-food Canada, Ottawa, Ontario, Canada.

ELLIOT, Royce and R.D. ANDERSON (1995), "Living with the high country: looking past the end of the Rabbit and Land Management Program", New Zealand Ministry of Agriculture and Fisheries, Wellington, New Zealand.

ENDICOT, Eve, ed. (1993), *Land Conservation through Public/Private Partnerships,* Island Press for Lincoln Institute of Land Policy, Covelo, California, United States of America.

FEATHER, Peter M. and Joseph COOPER (1995), "Voluntary incentives for reducing agricultural nonpoint source water pollution", *Agriculture Information Bulletin* No. 716, May, ERS-NASS, Herndon, Virginia, United States of America.

GAO [United States General Accounting Office] (1995), *Agriculture and the Environment – Information on and Characteristics of Selected Watershed Projects,* Report No. GAO/RCED-95-218, GAO Document Distribution Center, Gaithersburg, Maryland, United States of America.

GLACHANT, Matthieu (1994), "The setting of voluntary agreements between industry and government: bargaining and efficiency", *Business Strategy and the Environment,* Vol. 3, No. 2, pp. 43-49.

GLACHANT, Matthieu (1996), "The cost efficiency of voluntary agreements for regulating industrial pollution: a Coasean approach", paper presented at the *International Conference on the Economics and Law of Voluntary Approaches in Environmental Policy,* Venice, 18-19 November 1996, Cerna, Ecole de Mines de Paris, France.

GORRIE, Geoff (1995), "A case study: 'Landcare in Australia'", address to the April 1995 meeting of the Commission for Sustainable Development (mimeograph), Land Resources Division, Department of Primary Industries and Energy, Canberra, ACT, Australia.

GOSS, Kevin (1993), "The achievements of Landcare: a national and state perspective", paper presented to the *Outlook '93* conference, Canberra, ACT, Australia, 2-4 February 1993.

GOSS, Kevin, Tony CHISHOLM, Dean GRAETZ, Ian NOBLE and Michele BARSON (1995), "State of the agricultural resource base: a framework for government response", in *Sustaining the Agricultural Resource Base* (K. Goss et al., eds.), pp. 7-17, AGPS Press, Canberra ACT, Australia.

Government of Canada (1990), *Canada's Green Plan for a Healthy Environment,* Minister of Supply and Services, Ottawa, Canada.

Government of Canada (1994), *Report of Canada to the United Nations Commission on Sustainable Development,* Third Session of the Commission, Minister of Supply and Services, Ottawa, Canada.

GRETTON, Paul and Umme SALMA (1996), *Land Degradation and the Australian Agricultural Industry,* Industry Commission Staff Information Paper, AGPS Press, Canberra ACT, Australia.

GROSSMAN, Margaret Rosso and Wim BRUSSAARD, eds. (1992), *Agrarian Land Law in the Western World,* CAB International, Wallingford, Oxon., United Kingdom.

HANNA, Susan (1995), "Efficiencies of user participation in natural resource management", in *Property Rights and the Environment – Social and Ecological Issues* (Susan Hanna and Mohan Munashinghe, eds.), Beijer International Institute of Ecological Economics and The World Bank, Washington, DC, United States of America.

HEES, Eric, Henk RENTING and Sabine de ROOIJ (1994), *Naar lokale zelfregulering – Samenwerkingsverbanden voor integratie van landbouw, milieu, natuur en landschap* (Towards local self-regulation – co-operatives for integration of agriculture, environment, nature and landscape), Landbouwuniversiteit Wageningen, Wageningen, The Netherlands.

HILTS, Stewart G. (1995), "The evolution of environmental enhancement programs for agriculture in Ontario", paper presented at the Conference on Environmental Enhancement through Agriculture, Boston, Massachussetts, 15-17 November 1995.

HYBERG, Bengt and Sean PASCOE (1991), "Australia's environmental degradation from agriculture: lingering effects and greater visibility", *World Agriculture*, No. 63, June, pp. 36-39.

International Energy Agency (1995), *1995 IEA Survey on Voluntary Approaches*, OECD, Paris.

KNOPKE, P. and J.M. HARRIS (1991), "Changes in input use on Australian farms", *Agriculture and Resources Quarterly*, Vol. 3, No. 2, pp. 230-240.

Landcare Australia Ltd. (1994), *Fifth Annual Report*, Sydney.

LEMANN, Nicholas (1996), "Kicking in groups", *The Atlantic Monthly*, April, pp. 22-26.

LOEHMAN, Edna and Ariel DINAR (1994), "Co-operative solution of local externality problems: a case of mechanism design applied to irrigation", *Journal of Environmental Economics and Management*, Vol. 26, No. 3, May, pp. 235-256.

MARSHALL, Alfred (1952 [1920]), *Principles of Economics*, MacMillan, London.

MASUHR, Klaus P. (1994), *The External Costs of Energy Use – Internalization without the State?*, Bundesministerium für Wirtschaft, Bonn, Germany.

McNAIRN, Heather E. and Bruce MITCHELL (1992), "Locus of control and farmer orientation: effects on conservation adoption", *Journal of Agricultural and Environmental Ethics*, Vol. 5, No. 1, pp. 87-101.

Ministry for the Environment (1996), *Sustainable Land Management – A Strategy for New Zealand*, Ministry for the Environment, Wellington, New Zealand.

Ministry of Research, Science and Technology (1996), "A National Science Strategy for Sustainable Land Management", mimeograph, Ministry of Research, Science and Technology, Wellington, New Zealand.

MUES, Colin and D. COLLINS (1993), "A review of Commonwealth land care initiatives – promoting sustainable farming systems", paper presented to the *Outlook '93* conference, Canberra, ACT, Australia, 2-4 February 1993.

MUES, Colin, Heather ROPER and Jason OCKERBY (1994), *Survey of Landcare and Land Management Practices, 1992-93*, ABARE Research Report 94.6, Australian Bureau of Agricultural and Resource Economics, Canberra, ACT, Australia.

National Farmers' Federation, Australia (1991), "The role of farmer organisations and other non-governmental agencies in Landcare", in Proceedings of the International Conference on *The Environment and Sustainable Growth: The Key Role of Farmers*, 16-18 October 1991, Reykjavik, Iceland, pp. 210-218, International Federation of Agricultural Producers, Paris, France.

NELSON, Rohan A. and Colin MUES (1993), "A survey of Landcare in Australia", *Agriculture and Resources Quarterly*, Vol. 5, No. 3, pp. 400-410.

NEW, E. (1995), "Sustainable land management initiatives and implementation in New Zealand – Southland Sustainable Land Management Group", Background Paper 1A, New Zealand.

New York City Department of Environmental Protection (1990), "Discussion draft: proposed regulations for the protection from contamination, degradation and pollution of the New York City water supply and its sources", New York, United States of America.

NIELSEN, J.R. and T. VEDSMAND (1997), "Fisheries co-management: an alternative strategy in fisheries – cases from Denmark", in *Towards Sustainable Fisheries – Issue Papers* (free document), OECD, Paris.

OCKERBY, Jason (1995), "Regional salinity management", *Australian Commodities*, Vol. 2, No. 2, June, pp. 218-231.

OECD (1994), *Farm Employment and Economic Adjustment in OECD Countries*, Paris.

OECD (1995), *Environmental Performance Reviews – Netherlands*, Paris.

OECD (1997a), *Agriculture, Pesticides and the Environment – Policy Options*, Paris.

OECD (1997b), *Agricultural Policies in OECD Countries – Measurement of Support and Background Information 1997*, Paris.

OECD (1997c), *Towards Sustainable Fisheries – Economic Aspects of the Management of Living Marine Resources*, Paris.

OLSON, Mancur (1971), *The Logic of Collective Action: Public Goods and the Theory of Groups*, Harvard University Press, Cambridge, Massachusetts, United States of America.

OLSON, Mancur Jr. and Richard ZECKHAUSER (1970), "The efficient production of external economies", *The American Economic Review*, Vol. LX, pp. 512-517.

Ontario Farm Environmental Coalition (1994), *Environmental Farm Planning Workbook*, Toronto, Ontario, Canada.

PETERSON, Deborah (1995), "The land care taxation provisions – some issues", in *Outlook '95: Proceedings of the National Agricultural and Resources Outlook Conference, Canberra, 7-9 February 1995* (Vol. 1: Commodity Markets and Natural Resources, pp. 157-169), Australian Bureau of Agricultural and Resource Economics, Canberra, ACT, Australia.

POTIER, Michel (1994), "Agreement on the environment", *The OECD Observer*, No. 189, August/September, pp. 8-11.

President's Council on Sustainable Development (1996), *Sustainable America: A New Consensus*, US Government Printing Office, Washington, DC, United States of America.

PUTNAM, Robert D., Robert LEONARDI and Raffaella Y. NANETTI (1995), *Making Democracy Work: Civic Traditions in Modern Italy*, Princeton University Press, Princeton, N.J., United States of America.

RIDLEY, Matt (1996), *The Origins of Virtue*, Viking Press, Harmondsworth, Middlesex, United Kingdom.

ROBERTS, Brian (1995), "Are current incentives sufficient to encourage sustainable agriculture?", in Outlook '95: Proceedings of the National Agricultural and Resources Outlook Conference, Canberra, 7-9 February 1995 (Vol. I: Commodity Markets and Natural Resources, pp. 170-181), Australian Bureau of Agricultural and Resource Economics, Canberra, ACT, Australia.

ROUSSEAU, Jean-Jacques (1984 [1755]), A Discourse on Inequality, translated by Maurice Cranston, Penguin Books, Harmondsworth, UK.

RUNGE, Carlisle Ford (1984), "Institutions and the free rider: the assurance problem in collective action", The Journal of Politics, Vol. 46, pp. 154-181.

SINHA, Saurabh (1996), "The conditions for collective action: land tenure and farmers' groups in the Rajasthan Canal Project", Gatekeeper Series, No. 57, International Institute for Environment and Development, University of Sussex, Brighton, United Kingdom.

SMITH, Robert J. (1988), "Private solutions to conservation problems", in The Theory of Market Failure: A Critical Examination (Tyler Cohen, ed.), George Mason University Press, Fairfax, Virginia, United States of America.

SOLSBERY, Lee and Peter WIEDERKEHR (1995), "Voluntary approaches for energy-related CO_2 abatement", The OECD Observer, No. 196, October/November, pp. 41-45.

SRMC [Sustainable Resource Management Committee] (1995), "Evaluation report on the national decade of landcare plan", Draft Annex A, Canberra, ACT, Australia.

STRANLUND, John K. (1995), "Public mechanisms to support compliance to an environmental norm", Journal of Environmental Economics and Management, Vol. 27, No. 2, March, pp. 205-222.

van BROEKHUIZEN, R., L. KLEP, H. OOSTINDIE and J.D. van DER PLOEG, eds. (1997), Atlas van het Verniuwend Platteland, [Atlas of the Renewed Countryside], Misset Uitgeverij, Doetinchem, The Netherlands.

VANCLAY, Frank M. (1986), "Socio-economic correlates of adoption of soil conservation technology", Masters of Social Science thesis, University of Queensland, St. Lucia, Australia.

VANCLAY, Frank M. (1992), "The social context of farmers' adoption of environmentally sound farming practices", in Agriculture, Environment and Society (G. Lawrence, F. Vanclay and B. Furze, eds.), Macmillan, Melbourne, Australia.

VASSILIOU, Agapi and Gert van DIJK (1995), "The role of environmental co-operatives in the market of emission rights: the case of The Netherlands", in Environmental and Land Use Issues: An Economic Perspective, Proceedings of the 34th EAAE Seminar held in the Mediterranean Agronomic Institute of Zaragoza, 7-9 February 1994, Zaragoza, Spain (L.M. Albisu and C. Romero, eds.), pp. 504-516, Wissenschaftsverlag, Vauk, Kiel, Germany.

WALL, G.J., E.A. PRINGLE, G.A. PADBURY, H.W. REES, J. TAJEK, L.J.P. van VLIET, C.T. STUSHNOFF, R.G. EILERS and J.M. COSETTE (1995), "Erosion", in The Health of our Soils – Toward Sustainable Agriculture in Canada (D.F. Acton and L.J. Gregorich, eds.), pp. 61-76, Agriculture and Agri-food Canada, Ottawa, Ontario, Canada.

WEAVER, Robert D. (1996), "Prosocial behavior: private contributions to agriculture's impact on the environment", *Land Economics*, Vol. 72, No. 2, pp. 231-47.

WILLIAMS, John, Keith R. HELYAR, Richard S.B. GREENE and Rosemary A. HOOK (1993), "Soil characteristics and processes critical to the sustainable use of grasslands in arid, semi-arid and seasonally dry environments", in *Proceedings of the XVII International Grassland Congress*, 8-21 February 1993, New Zealand, pp. 1335-1350.

WOODALL, Peter (1996), "Farm films recovery scheme breaks new ground in 1995", Press Release No. 96/9651/3, 24 January 1996, Public Relations Principles Limited on behalf of the FFPG, Nottingham, United Kingdom.

WOODALL, Peter (1997), "Nationwide farm films collection scheme suspended – pioneering industry scheme receives no legislative support", Press Release No. 97/0716/3, 4 February 1997, Public Relations Principles Limited on behalf of the FFPG, Nottingham, United Kingdom.

WRR [Netherlands Scientific Council for Government Policy] (1992), *Environmental policy: strategy, instruments and enforcement,* English summary of the 41st report, Wetenschapelijk Raad voor Regeringsonderzoek, Den Haag, The Netherlands.

OECD PUBLICATIONS, 2, rue André-Pascal, 75775 PARIS CEDEX 16
PRINTED IN FRANCE
(51 98 05 1 P) ISBN 92-64-15418-3 – No. 49259 1998